8.95

D1236600

THE WORLD OF GOLD TODAY

✳

THE
WORLD
OF
GOLD
TODAY
by
Timothy Green

WALKER AND COMPANY • NEW YORK

First published in the United States of America
in 1973 by the Walker Publishing Company, Inc.

Published simultaneously in Canada
by Fitzhenry & Whiteside, Limited, Toronto.

ISBN: 0-8027-0437-9

Library of Congress Catalog Card Number: 73-83313

Printed in the United States of America.

To Snowball and Miranda,
who put up with my being away.

Contents

Preface

This book and my own interest in gold originally stem from an article I wrote for the American business magazine *Fortune* on the London gold market and the private buying of gold. My researches for *Fortune* led me into the little-explored world of gold dealing and gold smuggling, which I found vastly intriguing. To most of us in Britain or the United States gold is now rather a remote metal, which we assume lies in the vaults of banks. In fact most of the newly mined gold in the world today is going into the hands of private buyers.

This book looks at this world to see who produces gold, who markets it, who buys it, who smuggles it. Unlike the majority of writers on gold, I am not an economist and I am not putting forward my theories about the role gold should or should not play in our international monetary system. I have regarded my work to be simply that of a journalist investigating and reporting on the world of gold.

None of these investigations would have started without the help and encouragement of the editors of *Fortune*, in giving me the initial assignment on gold. I am very grateful to them and particularly to Murray Gart, now chief of correspondents for *Time*, for much guidance at the outset.

The book was first published in March 1968 at the peak of the international gold crisis. While it was thus topical, it was also speedily made outdated by the creation of the two-tier gold market. I have now, therefore, re-written it to take account of the many changes on the gold scene in the last five years. Many bullion dealers and mining executives have again been kind enough to give up time to talk to me and I appreciate their continuing courtesy. David Lloyd-Jacob and Peter Fells at Consolidated Gold Fields have also enabled me to expand my own knowledge of gold through the work I have done for them on their gold market surveys. I am grateful for their permission to reprint some of the Gold Fields statistical charts at the end of the book. Any errors and misinterpretations, however, are entirely of my own making.

All monetary statistics throughout the book are quoted in U.S. dollars. This is normal practice in the gold business, where even the official London price has been fixed in dollars since 1968; it also saves the confusion of switching back and forth from one currency to another in a book discussing many nations and currencies. The exceptions to this rule are in Chapters 1 and 2 when I have mentioned the changing values of gold throughout history. There prices are in the currency of the country at the time.

$1 billion means one thousand million dollars.

Since the gold business weighs in troy ounces, all references to ounces are on this scale. One troy ounce equals 1·1 ounces avoirdupois or 31·1035 grams. Where reference is made to tons, this is metric tons.

Finally, I must again thank my wife for her painstaking editing and Angela Cox who has typed the book with precision and speed.

T.S.G.

Dulwich

The World of Gold

A million dollars' worth of gold looks no larger than a puffed up yellow cushion on a footstool. Just eighty bars, each less than a foot long, gleaming on a wooden pallet, the whole pile less than 2 feet high. It almost seems like a hoax. Yet, as Disraeli once told the House of Commons, more men have been knocked off balance by gold than by love. For over 6,000 years men—and women—have fought for it, died for it, cheated for it, slaved for it. 'Get gold,' wrote King Ferdinand of Spain to his men in South America in 1511, 'humanely if you can, but at all hazards get gold.'

The civilisations of ancient Egypt and of Rome were nourished by gold, wrested from mines in conditions of unbelievable misery. 'There is absolutely no consideration nor relaxation for sick or maimed, for aged man or weak woman,' wrote the historian Diodorus in the second century B.C. 'All are forced to labour at their tasks until they die, worn out by misery amid their toil.'

The grip of gold has not been shaken off in the space age—indeed the umbilical cord that tethered astronaut Edward White, the first American to walk in space, to his Gemini spacecraft was gold plated to reflect thermal radiation. When men landed on the moon, gold foil shrouded the miniature television camera and other instruments on the moon buggies to protect them from the fierce rays of the sun in space. Back on earth, gold had its great champion throughout the 1960s in Charles de Gaulle. As he declared at a famous press conference in 1965, 'There can be no other criterion, no other standard than gold. Yes, gold which never changes, which can be shaped into ingots, bars, coins, which has no nationality and which is eternally and universally accepted as the unalterable fiduciary value par excellence.'

Logically, of course, the grip of gold is absurd, as Professor Robert Triffin of Yale University has pointed out: 'Nobody could ever have conceived of a more absurd waste of human resources than to dig gold in distant corners of the earth for the sole purpose of transporting it and reburying it immediately afterward in other deep holes, especially excavated to receive it and heavily guarded to protect it.'[1] But where gold is concerned, emotion rather than logic tends to prevail.

What John Maynard Keynes called 'this barbarous relic' still clings tenaciously to men's hearts. It remains the only universally accepted medium of exchange, the ultimate method by which one nation, whether capitalist or communist, settles its debts with another. In time of war it can be a crucial buttress of a nation's fate. Italy in 1936 made a desperate appeal to all its women to turn in their gold wedding rings to the government to help pay for the war in Abyssinia. India made similar pleas to her gold-hoarding millions when faced with the Chinese invasion in 1963 and the later conflicts with Pakistan.

The importance bankers attach to gold as an essential bastion of a nation's wealth is more than equalled by insatiable individuals the world over who see gold as a sheet anchor against devaluation, insecure currencies and the hazards of war. The devaluation of sterling in November 1967 and consequent uncertainties about the dollar the following spring triggered off a rush into gold that threa-

[1] Robert Triffin, *Gold and the Dollar Crisis,* Yale University Press, New Haven, 1961.

tened to sap the entire monetary stocks of the western world; three billion dollars worth of gold was snapped up in a matter of weeks. In South Vietnam the gold market expanded in the late 1960s in tandem with the escalation of the war there, and quietened down again only with the American withdrawal. In the Middle East the weeks immediately before and the first three days of the Arab–Israeli war in June 1967 saw heavy gold buying by Arabs hedging against the effects the war might have on their local currencies. In India more than $300 million in gold is smuggled past the customs each year to satisfy a craving for gold built up over generations by religious and social customs. 'Gold,' says a London bullion dealer with thirty years' experience, 'is bedrock.'

Even in the supposedly sophisticated economies of Britain and the United States, surrounded by cheque books and credit cards, we still pay homage to gold. After all the ultimate accolade our society accords a pop singer is a gold record (actually it is only gold plated and a 45 rpm gold record contains a mere 0·03 grams of gold) when a disc sells a million copies. Bing Crosby has won the most; Elvis Presley and the Beatles are coming up fast.

What makes gold the noblest of metals? Its greatest strength is its indestructibility. Unlike silver it does not tarnish and it is not corroded by acid—except by a mixture of nitric and hydrochloric acid. Gold coins have been recovered from sunken treasure ships after two centuries beneath the sea, looking as bright as new. The finest achievements of the goldsmith's art in ancient Egypt and in Mycenae have been uncovered almost unscathed by archaeologists. Heinrich Schliemann, the great German archaeologist, who made most of his money in the California gold rush before going to excavate Troy and Mycenae, was able to telegraph the King of Greece, 'I have looked upon the face of Agamemnon,' after he had recovered a golden death mask at Mycenae. As for the golden treasures of Troy, which he dug up untarnished, they were taken from the Schliemann Museum in Berlin in 1945 by the Russians and have never been seen since. They may well have been melted down, after 3,000 years, into perfectly acceptable gold bars.

To primitive man the first appeal was obviously aesthetic. It gleamed warmly at him from the beds of streams and he found it easy to work. Its beauty and versatility swiftly recommended it above all other metals. It was almost as soft as putty, so malleable that it could be hammered cold by even a primitive goldsmith until it was a thin translucent wafer only five-millionths of an inch thick. One ounce of gold can be beaten into a sheet covering nearly 100 square feet. It is also so ductile that 1 ounce can be drawn into 50 miles of thin gold wire or used to plate a thread of copper or silver wire 1,000 miles long. Blended with differing amounts of silver (which is always found with gold), platinum and palladium, gold comes in white, red, yellow and green hues. It is such an excellent conductor of electricity that a microscopic circuit of liquid gold 'printed' on a strip of plastic can replace miles of wiring in a computer

It is so valuable that one strong gold smuggler can carry $50,000 of it beneath his shirt, slotted into pockets in a special canvas waistcoat. Gold refineries find it pays to shut down once a year to spring-clean their chimneys and rooftops to extract from the congealed soot and grime tiny particles of gold that have escaped from the furnaces. One London merchant bank used to take up the wooden floor of its vault once in twenty years and burn it, to melt out specks of gold that had rubbed off the soft bullion bars over a generation.

Yet in volume it is so dense that all the 78,000 tons, estimated to have been mined between 4000 B.C. and the end of 1972, could be contained in a vault measuring only 60 feet on each side. The concert hall of Lincoln Centre in New York or the Royal Festival Hall in London could accommodate it comfortably and still leave room for a full orchestra, plus an audience in the front stalls. It could all be transported by one modern oil tanker—that is if Lloyd's would ever accept the insurance on a cargo worth $100 billion.

Gold's beauty, its scarcity and its almost mystical appeal as a symbol of power very quickly won it divine attributes. Ancient myths and legends cast it as the child of Zeus, a metal with which to adorn temples and to offer as appeasement to the gods. But the Golden Fleece that Jason and his Argonauts sought so assiduously was no more than a sheepskin, which was commonly used in ancient times in a stream containing alluvial gold to trap the fine specks of metal in the fast flowing water.

Part of the ancient science of alchemy, which was practiced from well before the birth of Christ until the mid-seventeenth century, was directed at finding or preparing the 'philosophers' stone,' which could turn base metals into gold and silver. In its heyday alchemy was practiced by kings Heraclius I of Byzantium, James IV of Scotland and the Emperor Rudolf II. Charles II of England had an alchemical laboratory built beneath the royal bedchamber with access by a private staircase. Chaucer devoted one Canterbury Tale—the 'Canon's Yeoman's Tale'—to the pursuit of the philosophers' stone. The Canon's Yeoman explained:

I seye, my lord can switch subtilitee
That al this ground on which we been ryding,
Til that we come to Caunterbury toun,
He could al clene turne it up-so-doun,
And pave it al of silver and of gold.

All alchemists started supremely confident that the magic formula was within their grasp, only to be finally disillusioned. Bernard of Treves in 1450 was sure he had concocted an infallible recipe for gold. He mixed 2,000 egg yolks with equal parts of olive oil and vitriol, then cooked this goulash on a slow fire for two weeks. All it did was poison his pigs.

Yet apart from its aesthetic appeal gold has no intrinsic value. It is hard to imagine being cast up on a desert island with anything more useless than gold. And, of course, many primitive societies, particularly in the Pacific, have managed very well with no gold at all. They simply adopted the sperm-whale-tooth standard, the boars-with-curved-tusks standard or the shell standard. While the United States decreed that 11.37 grains of gold equalled one dollar, the Solomon Islands preferred a standard where 500 porpoise teeth could buy one wife of good qualities. Lower down the scale one shell ring equalled one human head, one very good pig, or one male slave of medium qualities. Samoa was happier with the mat standard. 'No lover of money was ever fonder of gold than a Samoan was of his fine mats,' wrote one historian of the Pacific Islands.[1]

There is equal logic—or perhaps lack of it—in the hoarding of mats and gold. As the economist Paul Einzig pointed out, 'The production of shell money in the Pacific for the sake of being piled up in the house of a chief . . . is neither more futile nor less futile than the labour spent on the mining of gold for the sake of being able to bury it once more in the vaults of Fort Knox.'[2]

In fact, for all the trumpetings about gold, it has been available in any real quantity only for the last one hundred and twenty years. The true gold standard existed only in

[1] George Turner, *Samoa a Hundred Years Ago,* London 1884.
[2] Paul Einzig, *Primitive Money,* Eyre and Spottiswoode, London 1948.

the fifty years preceding the First World War. Before the California rush of 1848–49 ushered in an 'Age of Gold,' the metal was in very short supply. Indeed, some calculations suggested that up to 1850 only about 10,000 tons of gold had been mined since the beginning of time. There simply would not have been enough to implement a gold standard. It had been used simply as a commodity valuable for ornamental purposes and as a store of wealth for kings, princes, the Church and rich merchants. True, Croesus, King of Lydia (western Turkey), is credited with ordering the world's first gold coins to be struck as far back as 550 B.C., and the Greeks and Romans also made extensive use of gold coins. However, silver and copper coins were used in harness with gold—and were much more widely used as a medium of exchange between ordinary individuals. After all, only an infinitesimal amount of gold is needed to buy a loaf of bread. Silver was also much more widely employed as a standard of value. Until the nineteenth century, in fact, most countries were on either a silver standard or a bimetallic system of silver and gold, with the two metals in a ratio of 15:1 or 16:1.

It was only the increasing supplies of gold, first from Brazil in the eighteenth century, then from Russia, California, Australia and South Africa in the nineteenth century, that led to the demonetisation of silver and enabled gold to become the sole standard of value. More gold was mined in the hundred years 1800–1900 than in the preceding five thousand.

This sudden abundance of gold led Britain to go onto a formal gold standard in 1816—although she had been on one effectively since 1717, when Sir Isaac Newton, as Master of the Mint, established a fixed price of gold at £4 4s. 11½d. per troy ounce. The rest of Europe followed only in the 1870s. The United States, however, did not finally divorce itself from a bimetallic system of silver and gold until 1900—about the same time as India. China never made the switch from silver to gold. And the great age of the gold standard was all over essentially by 1914. Some economists, who now hail the gold standard as the only answer to the ills of the international monetary system, conveniently forget that the moment the gold standard faced its first real test, in World War I, it was almost universally suspended.

The ground rules of the old gold standard were a fixed price for gold, with gold coin forming either the whole circulation of currency within a country or circulating with notes representing and redeemable in gold. On the international plane it meant a completely free import and export of gold, with all balance of payments deficits settled in gold. Thus, in theory at least, gold disciplined the economy of a country. If it was running a balance of payments deficit, gold flowed out so there was less available for internal circulation; thus prices were controlled or came down, exports became more competitive and the balance of payments improved. On the other hand if a nation had a favourable balance of payments, gold came pouring in and the economy expanded. This was the standard which Britain adopted unconsciously in 1717 (although silver was not demonetised, it was used very little in the eighteenth century) and formally in 1816 at the end of the Napoleonic wars. She came off it unofficially in 1914, officially in 1919, never to return. The United States stayed with gold until March 1933, with a brief suspension from 1917 to 1919.

At the insistence of Winston Churchill, then Chancellor of the Exchequer, Britain did stagger onto a gold bullion standard in 1925. The main difference between the gold standard and the gold bullion standard was that gold coins did not circulate freely internally and could not be exchanged for other coin; notes could, however, be redeemed for 400-ounce bullion bars by anyone wishing or able to pay out $8,000 in notes. The gold bullion standard survived, however, for only five years.

Since the 1930s all that has remained is the international gold exchange standard, which prevailed until August 1972, when that too was put into limbo. The exchange standard meant, quite simply, that central banks supplemented their own gold reserves with certain key currencies (basically dollars) which could be redeemed for gold. Before World War II these reserve currencies amounted to little more than 10 per cent of the world's monetary reserves. Today, however, they form a much larger part than gold; in 1972 monetary gold accounted for only $45 billion of non-communist reserves of $137 billion—currencies and the new Special Drawing Rights provided the rest. The SDRs, sometimes known as paper gold, were first issued by the International Monetary Fund in 1970 as an initial step in creating a new reserve unit not linked to any specific currency. So far the SDRs represent a very small part of reserves, but their very existence is a significant step along the road that is gradually by-passing gold as a monetary metal.

At the moment the great debate is precisely how the by-pass should be designed. At its annual meeting in 1972, the IMF appointed a special committee of twenty under the Bank of England's Jeremy Morse, to consider how the international monetary system should be revised. The future role of gold will be a major point to be determined. There is, of course, a strong monetary gold lobby that cries constantly for a return to the good old gold standard as the cure-all for the world's monetary troubles. But they are really trying to reverse the whole history of the twentieth century which has seen gold shuffled more and more to the side-lines of the monetary system. Its latest setback came on 15 August 1971 when the Americans refused to go on paying out gold in exchange for dollars offered them by authorised central banks. They closed the famous 'gold window', probably for the last time. So the old phrase that 'the dollar is as good as gold', which prevailed under the gold exchange standard, is out of date. The closing of the gold window was closely followed by the devaluation of the dollar, bringing about the first change in the monetary price of gold since 1934. The Smithsonian Agreement signed in Washington in December 1971 raised the price from $35 an ounce to $38 —a very modest increase after thirty-eight years, if one considers how other prices have soared in the meantime. And, as it turned out, it was also an insufficient one. Barely fourteen months later, on 13 February 1973, the dollar was devalued by a further ten per cent, thus pushing up the monetary price to $42·22. Two such rapid changes in the monetary level are unprecedented. Previously an increase had been rare phenomenon; the price set by Sir Isaac Newton as Master of the Mint in 1717 lasted two hundred years; the changes in 1971 and 1973 were only the third and fourth since then. But they were so modest that it is highly debatable whether the jump from $35 to $42·22 will suffice as long as previous ones. With the free market price by comparison at well over $60 an ounce for much of 1972 and pressing on to over $100 in May 1973, the monetary price hardly looks realistic.

For most private individuals, of course, gold is now remote as a monetary metal; they have got quite accustomed to paper money which is completely divorced from gold. Keynes christened this the change from the age of Commodity Money to the age of Representative Money. 'Gold has ceased to be a coin, a hard, tangible claim to wealth, of which the value cannot slip away so long as the hand of the individual clutches the material stuff,' he wrote. 'It has become a much more abstract thing—just a

standard of value.'[1] In some nations this Representative
Money—the domestic note issue—is backed by a certain
percentage of gold, even though the notes themselves
cannot be redeemed for gold. But in Britain the domestic
note issue has been effectively divorced from gold since
1939, when the gold earmarked as backing for notes was
transferred to the Exchange Equalisation Account to meet
overseas demands. The United States required 25 per cent
gold backing for Federal Reserve notes until the spring
of 1968, when Congress issued legislation removing this
obligation—thus freeing more than $10 billion in gold to
meet overseas demands. But in both countries the private
citizen is forbidden to possess gold. An American citizen is
not permitted to hold gold at home or abroad, nor is a
British resident (the hairsplitting here means that a British
citizen residing permanently in France, or any other nation
that does permit individuals to hold gold, can do so).[2]

This remoteness prompted Professor D. H. Robertson in
his *Essays on Monetary Theory* to devise a conversation between
Socrates and a modern economist.[3]

SOCRATES: I see that your chief piece of money carries a
legend affirming it is a promise to pay the bearer the
sum of one pound. What is this thing, a pound, of which
payment is thus promised?

ECONOMIST: A pound is the British unit of account.

SOCRATES: So there is, I suppose, some concrete object
which embodies more firmly that abstract unit of account
than does this paper promise?

ECONOMIST: There is no such object, O Socrates!

[1] John Maynard Keynes, *Treatise on Money*, Vol. II, Macmillan, London,
1930.
[2] In April 1973 the U.S. Senate passed a bill approving private gold
buying.
[3] Professor D. H. Robertson, *Essays in Monetary Theory*, P. S. King & Son,
London, 1940.

SOCRATES: Indeed? Then what your Bank promises is to give me another promise stamped with a different number in case I should regard the number stamped on this promise as in some way ill-omened?

ECONOMIST: It would seem, indeed, to be promising something of that kind.

Although we may now be witnessing the final chapter in the history of gold as a monetary metal, the end is not yet in sight and, in best Agatha Christie traditions, the denouement is unknown. Attitudes to gold still vary too much from continent to continent for there to be any sudden curtain. Sit in an air conditioned banker's office just off Wall Street in New York and a smart-suited young man straight out of the Harvard Business School will say, 'The problem is how to get the demonetisation of gold.' And an official of the U.S. Treasury in Washington will back him up by saying, 'Try to sell an American a $5 gold piece and he wouldn't know what to do with it.'

Over in Europe, it is a very different story. In Paris economist Jacques Rueff still argues tirelessly that the world is on the brink of monetary catastrophe unless more attention is paid to gold, and the old disciplines of the gold standard restored. Rueff wants to double the price and insists all international payments should be made in gold. Rueff has been a part of the French financial scene since the mid-1920s, when he was a financial adviser to Poincare. He is a former deputy governor of the Bank of France and can certainly claim to have had practical experience of the mysteries of gold and gold exchange standards for longer than almost any other economist. The French, as the world's greatest gold hoarders, obviously agree with him.

Across the Alps, the Swiss bankers are also great champions of gold. Dr. Samuel Schweizer, former chairman of the Swiss Bank Corporation, argues, 'If you look at it from a realistic point of view you will appreciate that it is quite impossible, for a long time to come, to dispense with gold. Gold has such a deep foundation in the beliefs of the people of the world.' Much of that foundation is tucked away in the vaults of Swiss banks, where many of the world's private buyers keep their gold safely stored against evil days.

The fascination with gold continues in South Africa, which produces nearly 80 per cent of the noncommunist world's gold. Here Dr. William J. Busschau, an economist and former chairman of Gold Fields of South Africa, is one of the acknowledged experts of the world of gold. 'I decided in 1931,' he once told me, 'that in twenty years I wanted to know more about gold than any other man in the world.' His intensive studies made him one of the great champions of gold: 'The gold standard may not be perfect, but it's the best international system humanity has devised. And remember, gold isn't like peas that go off by the evening—it lasts forever.'

Across the Indian Ocean in India the grip of gold is just as strong. Walk slowly through the narrow jumble of alleyways of the Javeri Bazaar in Bombay, sidestepping the sacred cows, and there are scores of little goldsmiths' shops hung with a jumble of 22-carat necklaces, bracelets, armlets, anklets. Gold to the Indian is like an American Express card and his life insurance policy to an American. It is as much a part of the way of life as the caste system and the sacred cows. It will take a social revolution to break the hold of all three.

Farther east in Hong Kong, stand in the rich, evening sunlight on the tip of Hong Kong island and gaze across the South China Sea dotted with the sails of junks—almost all of them are carrying gold amid their cargoes of king-sized crabs and rice. For Hong Kong and Macao are the centres of a curious gold market that treads a delicate path between the legal and illegal in the world of gold. The junks and the planes from Hong Kong spirit the gold to every corner of the East. Hong Kong's streets may not be paved with gold, but the profits of the gold business have lined the walls of the Miramar night club with 300 ounces of gold, which gives off a seductive red-orange glow in the dim lighting. Its owner, who also runs the largest chain of jewellery stores and money exchange offices in Hong Kong, has no qualms about the future of gold. 'The appeal of gold,' he says contentedly, 'will last just as long as a gold ring can buy a family thirty catties of rice—which is enough to keep them alive for a month. It is an insurance policy.'

In Japan there are even those who, like the medieval alchemists, still see gold as an elixir. A hotel just outside Tokyo has installed a solid gold bath, fashioned in the shape of a phoenix from 143 kilograms of gold. It charges visitors 1,000 yen (about $3) for a two-minute dip in this golden tub, which supposedly adds three years to one's life. The Japanese stand in line for hours at the weekend to await their turn.

The challenge of gold is irresistible. In Florida today a former small-building contractor called Kip Wagner is well on the way to becoming a millionaire because he gambled on dredging up gold from the wrecks of fifteen ships of a Spanish treasure fleet that went down just off Cape Kennedy, Florida, in 1715 with $14 million (at the eighteenth-century valuation) in gold, silver, emeralds and porcelain in their holds. The first gold coins Wagner found on the sea bed in 1964 were as untarnished as newly minted gold despite 250 years in salt water. The first part of the treasure trove that Wagner and his Real Eight, Inc. syndicate scooped up from the remains of the galleons was sold for $227,450 at Sotheby's Parke-Bernet Galleries in New York in February 1967. It included an assortment of doubloons and magnificent circular solid gold ingots weighing 106 ounces each from the Mexico City Mint. The pride of the collection was a gold whistle attached to an 11-foot gold chain—the personal badge of office of the lost fleet's commander, Captain-General Don Juan Esteban de Ubilla. It sold for $50,000.

The chance of a gold-prospecting holiday, while rather unlikely to yield the treasure trove the Real Eight, Inc. found, can tempt many from the more relaxing pastime of sunning and sipping wine by the Mediterranean. When a British travel agency, Minitrek Expeditions, advertised vacations taking people 'North Cape Gold Prospecting,' they had far more applications than they could conceivably handle. The expedition offered six days at a remote gold camp on the Karasjokka river in Finnmark at the northern tip of Norway. Potential prospectors were promised that 'a few days' panning should yield sufficient nuggets to make a gold ring.'

Not even communism has yet been able to exorcise the power of gold, despite Lenin's fervent belief that its future use was limited to plating the walls of public lavatories. Russia, as the world's second largest producer of gold, has found it an invaluable means of exchange, enabling her to buy wheat and other vital commodities on the markets of the world. A Soviet official even conceded in 1972 that 'gold should be the basis of everything'. China seems to agree with him. The Chinese government wisely turned most of their sterling reserves into gold in 1965 and 1966 well ahead of the devaluation of sterling.

Perhaps they have all been reading George Bernard Shaw, whose widely quoted advice on gold was unequivocal. In *The Intelligent Woman's Guide to Socialism and Capitalism* he wrote, 'You have to choose (as a voter) between trusting to the natural stability of gold and the honesty and intelligence of members of the government. And, with due respect for these gentlemen, I advise you, as long as the capitalist system lasts, to vote for gold.'

2

History

'The United States is on the brink of an Age of Gold', said the *New York Herald Tribune* in November 1848 when the full scale of the gold discoveries in California began to percolate to New York. The newspaper might more correctly have phrased it that the world was on the brink. Suddenly in the nineteenth century the world of gold expanded beyond all previous understanding. The riches that Egypt had won nearly five thousand years before from the mines of Nubia, that the Roman Empire had wrested from Spain and that Spain herself had shipped from South America in the sixteenth century, were dwarfed by the avalanche of gold. In the short span of a hundred years, more gold was claimed from the earth than in the preceding five thousand. In the whole of the first century after Columbus discovered America the world output of gold totalled roughly 750 tons; in the last half of the nineteenth century it was a mighty 10,000 tons.

In London in the winter of 1852 members of the Banking Institute, trooping in from the cold February evening for their monthly meeting at 52 Threadneedle Street, were greeted by the warming spectacle of a 100-pound nugget of solid gold from Ballarat, Australia, resting on the chairman's table. Setting aside their top hats, they sat back to listen to a lecture by a Mr. Dalton on the impact of the gold discoveries. They learned to their astonishment that while the average annual value of new gold coins minted in

Britain, France and the United States had been $8.4 million before the new discoveries, it was more than $75 million in 1851.

The changes were not, however, simply a matter of cold statistics. In human terms the age of the individual prospector had arrived. Previously gold mining had always been the prerogative of, or at least had been heavily taxed by, the state. In the twentieth century it was to fall within the franchise of great mining corporations. Now, for a short span of half a century, the prospector—the man in his crumpled clothes and slouch hat, his ear constantly open at the diggings or in the saloon for the newest gossip of great finds—had his day. With his 'pan' he followed the latest rumours of gold—first to California, then to Australia, to New Zealand, back to Australia or Nevada, Colorado, Idaho or South Dakota, and finally in a glorious finale to the Klondike. 'The rush and struggle is awful', wrote one Australian gold digger, 'and the only chance is to fly off at the first sound. The mischief is that you hear many wonderful stories that prove false, that you will not listen to a first rumour, and by the time something authentic first reaches you, it is too late'.

California and the first gold rush of the common man did not, however, start the nineteenth century's Age of Gold. It was Russia, where landlords directed the toils of serfs, that led the world in gold until 1848. Far back in the history of gold, the rivers and streams of the Ural mountains had yielded rich deposits of alluvial gold, which had passed down the ancient trade routes to the Black Sea and the Mediterranean. And it was on the eastern slopes of the Urals that the resurgence of Russian gold mining began in 1744 with the discovery of a quartz outcrop near Ekaterinburg. The new mine was soon being run by the Czar and in its first forty years it produced 84,000 ounces. This relatively humble start in the eighteenth century stimulated the

Czars to greater exploration. In the first decades of the nineteenth century there were extensive searches for alluvial deposits in the Berezovsk region. Czar Alexander I, encouraged by his finance minister Kankrin, established in 1823 a commission of the heads of districts to take charge of the hunt for gold and to draw up regulations for the exploitation of deposits discovered. During the next seven years production in the Urals more than tripled from 50,000 ounces a year to 175,000 ounces, as many new alluvial gold fields were revealed in an area a hundred miles to the north and south of Ekaterinburg. This city became the centre of administration for the gold fields and all gold was hastened to the assay office there for analysis. Twice a year it was dispatched, under heavy guard, to St. Petersburg.

The success of these gold fields led to expeditions farther east in the Altai mountains and along the upper tributaries of the Yenisei River. By 1842 at least fifty-eight alluvial deposits were being worked in these remote regions of Siberia and the yield was 350,000 ounces. The hunt for gold was through rough, marshy terrain. A traveller wrote, 'One must have the iron constitution of the inhabitants of Siberia to bear such fatigue and privations; but even of them many succumb.' Only a small minority really benefited from the riches of the gold fields. The deposits were worked either directly for the Crown or by a handful of rich landlords, who were supposed to pay tax on their gold to the Czars. The labourers had to work every day except Sunday from 5 a.m. to 8 p.m. The only concession was that food was supposed to be available to them at fixed prices at the remote fields. Flogging was officially forbidden. The proprietors lived in great style. 'The Kalmyk in his felt cap brought me, on a plate of Japanese porcelain, oranges imported from Marseilles or Messina,' wrote one visitor to Siberia, who was entertained in regal fashion, 'whilst after a meal in which the delicacies of all climates had been brought under contribution, not forgetting the grape of Malaga, the Rhine and Bordeaux, came the aromatic nector of Arabia (coffee) along with excellent Havana cigars.' At the diggings the proprietors regaled themselves with champagne while watching their men at work.

Under these opulent landlords production soared until, by 1847, Russia was providing at least three-fifths of all the newly-mined gold in the world, most of it from the new gold-fields of eastern Siberia. Although Russian gold was rather eclipsed by the dramatic finds of California and Australia, exploration of new areas on the Lena River east of Lake Baikal and on the Amur River near the border with Mongolia kept output rising to 1·4 million ounces in 1880 and nearly 2 million ounces in 1914.

A by-product of this period of Russian expansion in gold was a short but spectacular flowering of the goldsmiths' art. In St. Petersburg Carl Fabergé was christened the new Cellini for his exotic golden masterpieces for the Czars and their families. Each Easter he made a golden Easter egg for the Czarinas. One year it was an exact model of a hen's egg; the outer shell was enamelled gold, it opened to reveal a golden yolk and this in turn opened to display a miniature hen. Inside the hen was a diamond reproduction of the Imperial Crown, inside this was a ruby pendant. To celebrate the opening of the Trans-Siberian Railroad in 1901, Fabergé's Easter egg contained a platinum engine which pulled golden railway carriages.

The champagne and Havana cigars of the Russian gentry and the golden Easter eggs of the Czarinas would definitely have seemed out of place in the California of 1848, although there was, of course, plenty to celebrate. California was to be the gold rush of the ordinary man, the simple labourer who threw up everything, trekked west and made his fortune. California also, unlike Russia, was to establish the gold fields as centres of rough and ready democracy where every miner, no matter what his origins, had his say and his vote.

It all began on a January afternoon in 1848, when a carpenter named James Marshall found what he thought were specks of gold in the tailrace of John Sutter's mill near the junction of the American and Sacramento rivers. One of Marshall's workmen recorded in his diary that night, 'This day some kind of mettle was found in the tail race that looks like gold, first discovered by James Martial, the Boss of the Mill.' In fact, Marshall was not at all sure that it was gold and he hurried back to Sutter's house to consult the *Encyclopedia Americana*. The description of gold in its pages so convinced him that he went rushing back to the mill in pouring rain and missed his supper. At first Marshall and Sutter tried to keep news of the discovery quiet, but rumours of gold were not easy to quench and soon the word had spread to San Francisco, then a struggling port of about 2,000 people. By spring half California had deserted its farms and homesteads and rushed to the gold fields. 'The whole country from San Francisco to Los Angeles and from the seashore to the base of the Sierra Nevadas resounds with the sordid cry of gold, GOLD, GOLD!' reported the *San Francisco Californian* in May 1848. 'The field is left half planted, the house half built, and everything neglected but the manufacture of shovels and pickaxes.'

The gold seekers were hardly disappointed. They found not only the sandbars and banks of the rivers near Sutter's Mill rich in alluvial gold, but were quickly able to trace deposits in other streams coming down from the western slopes of the High Sierras. Throughout that first summer the Californians had their find almost exclusively to themselves, for in the days before the telephone or cable the news of the discoveries percolated slowly even to the eastern seaboard of the United States. The 5,000 men working painstakingly along the riverbanks and up mountain streams panned out about a quarter of a million dollars' worth of gold—a promising beginning, but only one-fortieth of the yield of the following year.

To most Americans on the eastern seaboard California was still a remote, uncivilised strip of land which the United States was in the process of acquiring, along with New Mexico, from Mexico for $15 million. Indeed the ink seems barely to have been dry on the agreement with the Mexican government before the gold rush began. By the autumn of 1848 the first rumours of the discoveries were flying around New York. Each day brought fresh news and the excitement mounted. What happened during the next few months was quite unprecedented in history. Thousands of men suddenly saw a spark of opportunity to earn a fortune in a matter of days. Unlike previous gold discoveries, which had normally remained firmly under the prerogative of governments, here was a chance for anyone to stake his claim and dig riches. Even before President Polk finally confirmed the extent of the finds in a speech to Congress in December 1848, the scramble to get to the West Coast was on.

'It is well and truly said,' wrote a New York correspondent reporting to the *Banker's Magazine* in London, 'that the shoemaker is throwing away his last, the tailor his bodkin, the mason his trowel, the labourer his hod, the carpenter his chisel, the printer his stick, the painter his brush, the farmer his harrow, the quack his nostrums, the baker is leaving his dough, the butcher his stall, the clerk his desk and even the loafer his roost.'[1]

There were three routes to California: by ship around Cape Horn; by ship to Panama, then across the isthmus on a donkey and by ship on to San Francisco; or, finally, the long haul overland across the plains and through the mountains and deserts to the coast. Every transportation company, enjoying a heyday, pushed up its rates. First class

[1] *Banker's Magazine*, Vol. IX, 1849, p. 81.

by sea from New York to San Francisco cost $380, steerage was $200. A donkey across the Panama Isthmus was $30. Casualties along all the routes were high. Many prospective gold diggers failed to get over the fever-infested Chagres River on the Panama Isthmus. Cholera, Indians and accidents killed perhaps 5,000 of those taking the overland routes across the Great Plains in 1849.

For many of those who arrived early the price was worth it. Although there was an inevitable tendency for miners to exaggerate their finds, those first on the scene could earn anything from $300 to $500 a week—indeed a small fortune when most industrial workers in the eastern United States were earning only about $10 a week. The average wage at the 'diggins' in California in 1848 was as high as $20 per day, but this fell to about $6 a day by 1852. Yet, assuming there were 100,000 men actually working in the gold fields in that year, the total yield of $81 million in 1852 would suggest a very rough average of $800 for everyone—much more than incomes back in the East. Clearly, not everyone did make that much, but many men who went to California made at least a small fortune.

The thousands who rushed to California from all over the United States found Englishmen, Frenchmen and even the Chinese hard on their heels. This was an international gold rush. By mid-1849 one Scotsman was writing home to his family in Edinburgh 'This is the seventh week I have been here: we have averaged from 18 to 32 dollars every day till this week . . . as far as I can judge those who work steadily can make from 12 to 30 dollars per day. Cases are occurring of some getting from 100 to 200 dollars per day.'[1] By the end of 1849 there were at least 40,000 men actually working in the gold fields and they scoured out $10 million worth of gold.

Few of them knew the first thing about mining, but they

[1] *Ibid.*

quickly learned what telltale signs to look for amid the sand
and gravel in the bed of a creek and how to pan. It was
essentially a simple process. The miner filled his pan with
gravel and picked out the large stones by hand, then he
rotated the pan between his hands to keep the contents
suspended in the water. One side of the pan was tilted
slightly higher than the other so that the water carried away
the light particles while the heavier particles of gold were
left as a residue. As the search for gold became more
sophisticated the primitive panning process was supplemented by a 'cradle'—a long wooden box on rockers—
which enabled much larger quantities of gravel to be
handled at a time.

From the original find at Sutter's Mill, the miners ranged
out north and south along the Sacramento River and were
soon tracing the gold back into the Sierras. There the
prospectors soon located a belt of gold-bearing rock over
100 miles long and varying in width from a few hundred
feet to two miles. They called it the Mother Lode, for it was
from this quartz rock that the gold had been scoured over
the centuries and washed down the rivers. The mining
camps sprang up overnight wherever a promising new find
was located. The prospectors lived in leaky tents, lean-tos
or log cabins. It was a hard, often unrewarding, existence.
They toiled all day beneath the hot California sun up to
their waists in water. At night they went back to their huts
to eat whatever food might be available and to soak up bad
whisky. They got dysentery and scurvy. 'The miners of
California never shave; never put on clean vests, clean
dickeys or clean boots,' one traveller lamented. They were
indeed a scruffy lot, far from the comforts of wives and
families. The census of 1850 showed that 92·5 per cent of the
population was male. Most of the females were girls working
in the saloons amid their long gilt mirrors and red calico
curtains.

California was such a young country that it simply lacked the experience or the administrative machinery to cope with all the problems of the gold rush and a sudden population explosion of 100,000 newcomers in one year. The difficulties of law and order were hardly helped by the mingling of nationalities. There were 25,000 Frenchmen in California by 1853 and nearly 20,000 Chinese. Many of them did not speak a word of English. There was one ugly incident in Placerville when two Frenchmen and a Chinaman, caught in the act of robbery, were flogged by the miners, then tried and hanged, all without being able to utter a word in their defence. There were forty-seven illegal executions in California in 1855 compared with nine lawful ones. The murder rate in the gold-mining counties was terrifying; Calaveras County topped the list with thirty-two, El Dorado County claimed second place with twenty-six and Amador was third with twenty-one. The only real law was lynch law, which one Californian newspaper sadly admitted was better than no law at all. 'Lynch law is not the best law that might be, but it is better than none, and so far as benefit is derived from law, we have no other here.'

Throughout it all the gold production increased year by year to 2·5 million ounces in 1851, then a peak of 3 million in 1853. The United States Mint began coining Californian gold in such profusion that silver coins became scarce almost overnight. Across the Atlantic the gold reserves of the Bank of England rose from £12·8 million in 1848 to £20 million in 1852. 'As the creditor of the whole earth, London got the first of this gold,' notes Sir John Clapham in his *History of the Bank of England*.[1] France fared even better; the Bank of France's gold stock soared from £3·5 million in 1848 to £23·5 million four years later. The city editor of *The Times* suggested, 'There will be considerable surplus of gold; in

[1] Vol. II, p. 217, Cambridge University Press, Cambridge.

fact on the Continent, which is on a bimetallic system, more and more payments are likely to be made in gold.'

This was only the beginning. An Australian named Edward Hammond Hargraves, who had been to California, was certain that the same geological features were to be found in his own country. Returning on the boat from California late in 1850, he predicted he would find gold within a week. 'There's no gold in the country you're going to and if there is, that darned Queen of yours won't let you touch it,' a fellow passenger told him. 'There's as much gold in the country I'm going to as there is in California,' snapped Hargraves back, 'and Her Gracious Majesty the Queen, God bless her, will appoint me one of her Gold Commissioners.' Hargraves was right. Within one week of landing he had found gold on a tributary of the Macquarie River not far from Bathurst in New South Wales. The gold rush was on. 'A complete mental madness appears to have seized almost every member of the community,' the *Bathurst Free Press* reported. 'There has been a universal rush to the diggings.' Hargraves was duly made a Commissioner for Lands, received a reward of £10,000 plus a life pension. In 1854 he was presented at Court to Queen Victoria. For the rest of his life he was regarded as a man with the Midas touch. He was sent to look for gold in Tasmania and in Western Australia, but failed in both places.

The first news of the new gold field reached England, along with the first £800 worth of gold, aboard the *Thomas Arbuthnot*. Her captain said, 'The colony is completely paralysed. Every man and boy who is able to lift a shovel is off, or going off, to the diggings. Nearly every article of food has gone up, in some cases 200 per cent.' He had had to promise his crew double wages to get the boat away from Sydney; even then half a dozen had deserted. The impact of the Australian find on Britain was to be even more important than that of California. The bulk of the latter's gold stayed in the United States; 80 per cent of Australia's gold was to come through the London market. It is no coincidence that of the five members of the present London gold market, two were founded at this time. Pixley and Able (later to merge into Sharps, Pixley) was founded in 1852 by Stewart Pixley, formerly a senior clerk in the cashier's office at the Bank of England. Samuel Montagu's, the merchant bankers and bullion dealers, opened their door for the first time in 1853.

In fact Hargraves had touched only the fringe of Australian gold. New South Wales was yielding 850,000 ounces in 1852, but the neighbouring state of Victoria had joined the hunt. Victoria offered a reward of £200 for gold found within 200 miles of Melbourne. In the autumn of 1851, barely six months after the New South Wales discovery, gold was found at Ballarat, a mere 60 miles from Melbourne. Later the same year came another find at Bendigo Creek, 30 miles farther north. Now the flood gates were really open; 370,000 immigrants arrived in Australia in 1852. The colony, which a few years before had been peopled only by convicts who had been transported, and a handful of farmers, had its economy transformed forever. The Australian gold miners were never such a cosmopolitan bunch as their colleagues in California, but they insisted on the same democracy in the gold fields. There was, however, much more law and order in the Australian gold rush right from the start. Hard liquor was banned. Special gold commissioners were appointed to administer the diggings. They sold licences for 30 shillings a month and normally parcelled out 15 to 24 feet along a creek to a party of three to six men. They were a wildly assorted crowd. A visitor, G. L. Mundy, writing in 1852, reported, 'There were merchants, cabmen, magistrates and convicts, amateur gentlemen rocking the cradle merely to say they had done so, fashionable hairdressers and tailors, cooks, coachmen, lawyers' clerks and their masters, colliers, cobblers, quarrymen, doctors of physic and music, aldermen, an ADC on leave, scavengers, sailors, shorthand writers, a real live lord on his travels—all levelled by community of pursuit and of costume.' 'Levelled' is just the right word. The miners lived in bark huts or tents. 'Our furniture,' wrote one miner, James Bonwick, 'is of simple character. A box, a block of wood, or a bit of paling across a pail, serves as a table.' Meals were primitive. 'The chops can be picked out of the frying-pan, placed on a lump of bread, and cut with a clasp knife that has done good service in fossicking during the day.' Insects and flies added to the discomfort. 'The nuisance is the flies,' complained Bonwick, 'the little fly and the stinging monster March fly. O! the tortures these wretches give! In the hole, out of the hole, at meals or walking, it is all the same with these winged plagues. When washing at a waterhole, the March flies will settle upon the arms and face, and worry to that degree, that I have known men pitch their dishes, and stamp and growl with agony. The fleas, too, are of the Tom Thumb order of creation, and they begin their blood-thirsty work when the flies are tired of their recreation.'[1] For those who stuck it, the rewards could be handsome. Almost £11 million worth of gold was mined in Victoria alone in 1853; by 1856 it had risen to a peak of £12·25 million.

Yet the Australian rush, like that in California, had a relatively short life. In both countries the alluvial gold that was easily available on the surface was quickly scooped up. Once it had gone, the search for gold called for more patience and better equipment. By the mid 1850s the pattern of gold mining in both countries was changing. No longer was it the individual miner with his pan, but a group of men joining together and pooling their capital to build more elaborate crushing equipment and to dig

[1] James Bonwick, *Notes of a Gold Digger*, 1852.

deeper. As early as 1851 the San Francisco weekly *Alta California* noted, 'We have now the river bottoms and the quartz veins, but to get the gold from them we must employ gold.' Over $13·5 million was invested in California by 1859 in the building of deep mines and over 5,000 miles of canals, ditches and flumes were dug to supply water. The lean years had soon arrived. Average wages in California which had been $10 a day in 1850, had dropped to $3 ten years later.

Miners now dashed off at the least rumour of gold, seeking new prosperity. There were many false rumours in San Francisco newspapers specifically designed to dispose of surplus labour. There was a rush to the Fraser River in British Columbia in 1858, Pike's Peak, Colorado, in 1859 and to Boise, Idaho, in 1862. None of them produced gold on any scale compared to California. The richest discovery was the Comstock Lode near Virginia City, Nevada, in 1859. The Lode contained such high grade deposits of gold and silver that in the first twenty years in which it was mined it yielded $306 million of the two metals, including $130 million worth of gold. Among those in Virginia City at the height of the rush was Mark Twain, working for a spell as city editor of *The Enterprise*. He later wrote about the boom town in *Roughing It*. 'Money was as plentiful as dust: every individual considered himself wealthy, and a melancholy countenance was nowhere to be seen. The "city" of Virginia . . . claimed a population of fifteen to eighteen thousand, and all day long half of this little army swarmed the streets like bees and the other half swarmed among the drifts and tunnels of the "Comstock," hundreds of feet down in the earth directly under those same streets. Often we felt our chairs jar, and heard the faint boom of a blast down in the bowels of the earth under the office.'[1]

Across the Pacific, Australia was also having its secondary

[1] Mark Twain, *Roughing It*, Vol. II, Chatto and Windus, London.

rushes. One rumour of gold on Fitzroy River in Queensland in 1858 sent 10,000 people trailing north. It was a completely false lead, but because of lack of communications, there was no way of stemming the tide. Even New Zealand, which had jealously watched Australia's budding economy be nourished by gold, finally had its reward. Gold was found near Dunedin in the south island of New Zealand in 1861. More than 7,000 men were working the field by the following year and a steady production of 500,000 ounces a year was maintained until 1870.

But essentially this was all fits and starts and for almost thirty years from the mid-1850s until the late 1880s the flood of gold slowed down (although far more was still being produced every year than before 1848) and the gold that had been mined was gradually assimilated. When it restarted in South Africa, the character of the gold rush was to be radically changed. The first great riches found in South Africa were in fact not gold but diamonds. The diamond fields at Kimberley on the banks of the Vaal River were discovered in 1867. They attracted from all over the world men who had previously been lured by gold. Now they built up diamond fortunes that would later enable them to participate in the next scramble for gold. Among them were Cecil Rhodes and his partner Charles Rudd, J. B. Robinson, Hans Sauer, Alfred Beit, Hermann Eckstein, Lionel Phillips, Barney Barnato and George Albu, who were to establish the mining finance houses to nurture the South African gold mines.

During the early years of the Kimberley rush some gold was found in the Transvaal, primarily at Barberton in the east Transvaal in the shadow of the Drakensberg Mountains. It was never enough to tempt the diamond men of Kimberley. Lionel Phillips was sent down from Kimberley to view the new gold field, but he reported back quickly that the field had little potential. With Kimberley unimpressed the boom quickly petered out, although a few small mines continued to produce gold for many years. Yet prospectors were not discouraged. The problem was that they were looking for gold as it had appeared in California and Australia.

A. P. Cartwright, in *The Gold Miner*, says 'The prospectors of 1885 were slow, agonizingly slow, in their progress towards their unknown goal. They stumbled about as men do who are blind-folded, groping their way towards what they believed would be the "mother lode" from which had sprung the traces of gold they had found so far. They followed false trails, they panned in all the wrong places.'[1]

Yet all the time they were right on top of the richest gold field the world has ever known. They failed to understand the peculiar geology of the huge Witwatersrand basin in which the gold-bearing reefs, with gold flecks so fine that they could not normally be seen with the naked eye, out-cropped only briefly on the surface, near what is now Johannesburg, then plunged below the ground at an angle of 25 degrees or more, sloping inward toward the centre. The gold-bearing sides of the basin have never 'bottomed out.'

The credit for discovering the Main Reef of gold-bearing conglomerate—it looks like a sandwich of white pebbles packed tightly together—normally goes to a man named George Harrison, who, so the story goes, found the reef outcropping on Langlaagte farm in February 1886, when he was digging up stone to help build a house for Widow Oosthuizen, who owned the farm. It was hardly a sensational discovery like a find of alluvial gold. Harrison, who had had experience in the Australian gold fields, simply recognised the rock as a gold-bearing formation which, if crushed, might yield an ounce or two of gold from every ton of ore.

[1] A. P. Cartwright, *The Gold Miners*, Purnell & Sons (S.A.) Pty., Ltd., Cape Town, 1962.

This is the essence of the South African gold mines. No one picks up nuggets. There is unfathomable body of low grade ore stretching in the wide arc from 40 miles to the east of Johannesburg to 90 miles west, then swinging down southwest into the Orange Free State. The gold-bearing reefs, laid down perhaps 2,000 million years ago, vary in thickness from one-tenth of an inch to 100 feet, but on the average are only 1 foot thick. Except in rare outcrops they have been covered over the ages with thousands of feet of hard rock. Tracking them far below ground calls for rare skill in geological detective work; mining them calls for capital and engineering skill.

Thus, although the news of gold on Langlaagte farm brought men rushing to the fledgling city of Johannesburg, it was only those with capital who could participate. The diamond men from Kimberley quickly established control. They came discreetly up by coach, trying hard to avoid their rivals knowing where they were bound. J. B. Robinson and Hans Sauer happened to be riding the same coach, so at one stop each decided to leave the coach to try to prevent the other learning his real destination. There could, however be no real secret. Within two years four of the present seven mining finance houses had been established, all backed by men who had made their money in diamonds. The first was formed by Hermann Eckstein in 1887, was soon nicknamed 'The Corner House', and eventually became Rand Mines. Right behind came Cecil Rhodes and Charles Rudd with Gold Fields of South Africa, the Barnato brothers with Johannesburg Consolidated Investment Company, and George and Leopold Albu with General Mining and Finance Corporation. Adolf Goerz started a fifth group in 1893 after he came out to look over the gold fields for a group of Berlin businessmen. It eventually became the Union Corporation. But even the capital and mining experience of these men were not in themselves enough to get the South African gold-mining industry off to a flying start.

There were soon plenty of claims staked out all along the south fringe of Johannesburg wherever the Main Reef and its associated reefs of the Main Reef Leader, the Bird Reef and the South Reef outcropped, but the real problem was getting a high enough value of gold from the ore. The old methods of crushing ore to a powder which was then carried by water over copper plates coated with mercury, which amalgamated with the gold, might have been satisfactory for the gold in the quartz veins of California or Australia, but it was not subtle enough for the gold sprinkled finer than pepper throughout the Rand. They extracted at best 70 per cent of the gold, and on average 65 per cent. Assuming that each ton of ore contained 1 ounce of gold, this meant that only two-thirds of an ounce was being extracted—a loss of almost $4·20 a ton. Gold mining on those terms did not make sense and in 1890 the gold boom seemed finished. 'Grass will grow in the streets of Johannesburg within a year,' one miner predicted.

Indeed, it might have done so had not two doctors in Glasgow, Robert and William Forrest, and a chemist, John S. MacArthur, begun experimenting, quite independently, on the problem of gold extraction. In 1887 they patented the MacArthur–Forrest process for extracting the gold from ore by cyanide. Once the ore had been crushed to a fine powder it was circulated through tanks containing a weak solution of cyanide, which has an affinity for gold. The solution dissolved the gold (and silver) but had no effect on the rock particles (in the same way that if sugar and sand were stirred together in tea, the sugar would dissolve, the sand would remain as grains). The remaining rock pulp could thus be filtered off. Zinc dust was added to the cyanide solution and it replaced the gold, causing fine specks of gold to be precipitated out. The precipitate could then be refined. Properly developed and applied correctly, the MacArthur–Forrest process extracted 96 per cent of the gold from the ore. Without this invention the industry as we know it today would not exist. Although the actual equipment has been improved over the years, the MacArthur–Forrest technique is still the basis of gold extraction today.

While the MacArthur–Forrest process was being tentatively tried out, The Corner House was taking another major move forward. On the advice of an American mining engineer, J. S. Curtis, who had made a careful study of the angle at which the outcropping gold reefs plunged under-ground, they began to buy up, in great secrecy, large blocks of land to the south of the main outcrops. Boreholes soon revealed the gold reefs over 500 feet under ground. The Corner House, backed by the Rothschilds, then began the boom in the deep-level mines which have become such an essential part of the South African gold-mining industry. Even at this early stage the cost of establishing a new mine was considerable; at least £500,000 was needed. The yield of the mines quickly made such an investment worthwhile. In 1887 South Africa had contributed a mere 39,880 ounces of gold, a tiny ·8 per cent of world production. Just five years later her production topped the million-ounce mark, worth £4·5 million, and represented over 15 per cent of world production. By 1898, just before the Boer War shut down the gold industry for three years, output was up to almost 4 million ounces, more than a quarter of the world's newly mined gold. In that year she toppled the United States from first place in the world gold production league. Apart from the lull during the Boer War, South Africa has contributed more than a quarter of the new gold every year since 1898 and one-third each year since 1910. The total value of the gold mined in South Africa had reached over $35 billion by the end of 1972.

Just as the South African industry was getting into its stride, the individual prospector, who had been king since 1848, had a final glorious fling to bow out his century. Two prospectors, Robert Henderson and George Washington Carmack, were fishing for salmon on the Thron-diuck (it quickly became Klondike) tributary of the Yukon River in the far north of Canada one August afternoon in 1896 when the gleam of gold caught their eye. For several decades there had been wild rumours of gold in these streams of the far north, but few of them had lived up to expectation. The best source of gold had been in the creeks lower down the Yukon in the U.S. territory of Alaska, where a bustling community called Circle City had grown up with a music hall, two theatres, eight dance halls and no fewer than twenty-eight saloons. It was gaily christened 'The Paris of Alaska.' Henderson and Carmack's discovery on Thron-diuck made it a ghost town overnight. In that first autumn, as everyone from Circle City stampeded up the Yukon and the little community of Dawson city was born, the far north kept the secret to itself. Despite the desperate shortage of supplies (salt fetched its weight in gold), the prospectors held on against disease and starvation. A barber in Dawson City, got a small slice of one claim, went out and dug up $40,000 in gold. In Harry Ash's Saloon in Dawson City there was so much gold mingling with the sawdust on the floor that one enthusiast panned for gold right there. According to legend, he came up with $275 worth of gold dust that had filtered out of miners' pockets.

Came the spring of 1897 and the first packet steamers sailed south to Seattle and San Francisco laden with gold stuffed into buckskin bags, glass fruit jars, tomato tins and blankets tied with string. It was a tonic for which the West Coast had long been waiting. The heady excitement of California was long over and there had been little prosperity to follow it. Now in one final insane rush everyone was off to the Yukon. Fifteen hundred people sailed north from Seattle within ten days of the first news of gold. The mayor of that city, who was on a visit to San Francisco when the news came through, wired his resignation and raced north. Steamer offices were in a state of siege and tickets were selling for $1,000. By February of 1898 forty-one ships were on the regular run from San Francisco to Skagway, the nearest port to the gold field. From Skagway, the prospectors had to make the long haul up over the Chilkoot or the White Horse Pass and so down the Yukon to Dawson City. It was a harsh journey. Among the thousands who embarked on it many failed. Pierre Berton, in his excellent book on the Klondike, reckons that of the 100,000 who set out for Dawson City, 30,000 to 40,000 actually arrived. Of these, perhaps 5 000 searched for gold; a few hundred got rich.[1]

All along the way they fell victim to the weather and to men like Jefferson (Soapy) Smith, who ran the town of Skagway, cheating allcomers and killing any who argued. One horrified traveller wrote, 'I have stumbled upon a few tough corners of the globe but I think the most outrageously lawless quarter I ever struck was Skagway. It seemed as if the scum of the earth had hastened here to fleece, rob or murder. There was no law whatsoever; might was right, the dead shot only was immune to danger.' Skagway, of course, was on American soil, but to reach Dawson City the prospectors, tramping in a never-ending line up the snow-clad mountainsides, had to cross the Canadian border at the top of the pass. Here a handful of men from the Canadian Northwest Mounted Police sought to bring some sort of order to chaos. At gunpoint they refused to let through prospectors who were not carrying a year's supply of food. Major J. M. Walsh of the Northwest Mounted Police reported, 'Such a scene of havoc and destruction can

[1] Pierre Berton, *The Golden Trail*. Macmillan, Toronto 1954.

scarcely be imagined. Thousands of pack horses lie dead along the way, sometimes in bunches under cliffs, with pack saddles and packs where they have fallen from the rocks above.' Those who did get through swelled the population of Dawson City so fast that by the summer of 1898 it had become the largest Canadian city north of Winnipeg. It was a frenzied yet pathetic community, so short of supplies that the police would not bother to arrest a man unless he had his own provisions. Most boats that did come up the Yukon carried whisky instead of badly needed food. One year all the eggs were bad when they arrived and hungry citizens had to wait a year for a fresh supply. When the governess of the local bishop's children married a missionary the only thing he could find to give her for a present on her wedding day was a pot of marmalade.

It was all over as fast as it had begun. There was indeed plenty of gold in the creeks around Dawson and it continued to be worked on a commercial basis until the winter of 1966, but the horde of prospectors had picked the cream off the field by 1900. The Klondike rush probably yielded about 2·5 million ounces of gold in the last three years of what is certainly the most exciting century in the history of gold.

While the gold rushes of the nineteenth century had an immediate and profound effect in bringing vast hordes of people to virtually virgin territory and stimulating in a year faster growth of cities and communications than would otherwise have taken place in fifty, the impact was hardly less on the great financial centres of Europe. Not only was London to provide a major part of the capital to develop the South African gold-mining industry, but the gold itself flowed in. All South Africa's gold was refined in London by Rothschild's, Johnson Matthey and Raphael. Johnson Matthey did much of the early assay work on South African gold. They and Rothschild's played a major part in selling it through the London gold market.

The impact of Californian and Australian gold had quickly been reflected in the gold reserves of Britain, France and other major European powers. Inevitably gold was used more and more in business transactions, although at first only Britain was fairly and squarely on a gold standard. In 1871, however, Germany, getting fat on the French indemnity after the Franco-Prussian war, issued a new currency unit, the mark, which was based on gold. She bought £50 million worth of gold and coined it over two years. This demand for gold pushed up the price slightly and in relation to it the value of silver fell. More European countries were at once forced to switch to gold. Scandinavia demonetised silver in 1874; Holland followed a year later; France and Spain switched to gold in 1876. Before the end of the nineteenth century almost every country in the world had changed to a gold standard. Russia went onto gold in 1893; India adopted a gold exchange standard in 1898; and the United States capped it all by the Gold Standard Act of 1900.

The silver lobby in the United States had fought a desperate rear-guard action for almost half a century and had succeeded in keeping the country on a bimetallic system of gold and silver long after the flood of Californian gold had began. They had pushed through the Sherman Silver Purchase Act of 1890, which laid down that the government must buy 4·5 million ounces of silver a month with Treasury notes redeemable in gold or silver, 'it being the established policy of the United States to maintain the two metals on a parity with each other upon the present ratio' (16 : 1). But not for long. The 1896 Presidential election campaign, one of the fiercest ever fought, turned on the issue of silver. The Democrats, with the retention of bimetallism as the main plank of their platform, nominated William Jennings Bryan of Nebraska to fight the Republicans' William McKinley. In his speech to the

Democratic National Convention Bryan said: 'We will answer their demand for a gold standard by saying to them: You shall not press down upon the brow of labor this crown of thorns, you shall not crucify mankind upon a cross of gold.' All stirring stuff, but Bryan was defeated in the popular vote by 6·5 million to 7 million. Four years later the United States went quietly onto the gold standard. The dollar of 25 and four-fifths grains of gold 90 per cent fine became 'the standard unit of value; and all other forms of money issued or coined by the United States shall be maintained at a parity of value with this standard.'

When the world went to war in 1914, fifty-nine countries were on a gold or gold exchange standard; China was the sole major nation still on a silver standard. What was most remarkable was that the price of gold in 1914 was exactly the same as it had been in 1717, when Sir Isaac Newton as Master of the Mint had established a price of £4 4s. 11½d. per troy ounce. The price had remained stable so long partly because of the tremendous increase in gold production during the period. Throughout the whole eighteenth century gold production had totalled only 61 million ounces; one-quarter of that amount was mined in 1899 alone, and the total nineteenth century output was over 372 million ounces. By 1914 more than 20 million ounces of new gold were coming onto the market every year.

With the coming of war, however, most of Europe immediately went off the gold standard. Britain, in theory, kept full convertibility, but made the export of gold so difficult (aided by the U-boat blockade) that effectively the gold standard ceased to work. It was not until 1919 that the export of gold was officially prohibited and formal suspension of the gold standard became effective. The United States stopped the export of gold when she entered the war in 1917, but resumed full gold payments in 1919.

At a conference in Genoa in 1922 the whole future of the

international monetary scene was thrashed out by Europe's bankers. They decided that henceforward they would 'economise in the monetary use of gold through the maintenance of reserves in the form of balances in foreign currencies.' This was in fact the beginning of the gold exchange standard and throughout the 1920s banks began to keep an increasing part of their reserves in key currencies which could be exchanged for gold. Britain, however, made a partial return to the gold standard in 1925 at the insistence of Winston Churchill, then Chancellor of the Exchequer. More correctly, it was a gold bullion standard, for notes were no longer convertible into gold coin, but they could be exchanged for bullion in 400-ounce 'good delivery' bars. The old price of £4 4s. 11½d. per troy ounce was also reinstated—the price had gone as high as £5 18s. during the period off the gold standard.

Economists have never ceased to argue whether this return to the prewar price of gold when the standard was restored, was the major cause of its speedy collapse. It certainly seemed inconsistent with the general increase in prices that had taken place. The other great weakness of the 1920s was the lack of co-operation between central banks in trying to stablise prices and halt inflation.

The Wall Street crash of 1929, with its echoes throughout Europe, began an era of crisis for gold. Britain went off the gold bullion standard in 1931; sterling was devalued, creating a new price for gold that fluctuated between £5 10s. and £6. It was a move which caught several European central banks, particularly those of Belgium and Holland, in an embarrassing position, since they were holding large reserves of foreign currency instead of gold. It was hardly an auspicious start to the gold exchange standard and it has taken a long time to erase the memory of 1931 from the minds of many European bankers.

The United States stayed uneasily on the gold standard

until March 1933, when the newly inaugurated President Roosevelt, in an effort to give some sense of stability to U.S. banks, broke the dollar's link with gold by imposing a ban on the hoarding and export of gold. In the autumn of the same year Roosevelt, advised by Henry Morgenthau, then Acting Secretary of the Treasury, and Jesse Jones of the Reconstruction Finance Corporation, started fixing the price of gold each morning over breakfast, trying to push it up slightly in the hope it would pull the United States out of the depths of the Depression. Arthur M. Schlesinger, Jr., has described how 'While Roosevelt ate his eggs and drank his coffee, the group discussed what the day's price was to be. The precise figure each day was less important than the encouragement of a general upward trend. One day Morgenthau came in, more worried than usual, and suggested an increase from 19 to 22 cents. Roosevelt took one look at Morgenthau's anxious face and proposed 21 cents. "It's a lucky number," he said, with a laugh, "because it's three times seven." '[1]

The policy itself was not so lucky. At the end of January 1934 Roosevelt had to face the fact that the easing up of the price was not lifting the nation out of depression as had been expected. So the dollar price for gold was fixed at $35 an ounce, an increase of over $14 an ounce from its old price of $20·67 an ounce. Export of gold was permitted to recognised foreign central banks and government institutions; domestic hoarding remained forbidden.

The devaluation first of sterling, then of the dollar gave an enormous stimulus to the gold-mining industries of the world. It is ironic that throughout the harsh years of the 1930s these industries boomed as never before, urged on by the higher price for gold. During that decade world gold production doubled from 20 million ounces a year to a

---

[1] Arthur M. Schlesinger, Jr., *The Age of Roosevelt* (Vol. 2): *The Coming of the New Deal*, Houghton Miffin, Boston, Massachusetts, 1960.

peak of 40 million in 1940. In California thousands of the unemployed became gold prospectors, scouring the streams of the High Sierras for gold overlooked by the miners almost a hundred years earlier. United States gold production doubled during the 1930s and in 1940 reached a record total of just under 5 million ounces; Canada attained a peak of 5·5 million ounces in 1941. Neither country has come close to this record since. Indeed, throughout the thirties South Africa found herself being challenged as the premier gold-producing nation for the first and only time in the twentieth century. Her share of world production, which had climbed to over 50 per cent in the 1920s, suddenly pitched down to 32·5 per cent in 1938. But it was the final effort for her rivals. Their best ore reserves were spent. South Africa has had no further challenge, and after 1940 climbed into a more and more commanding position in the world of gold.

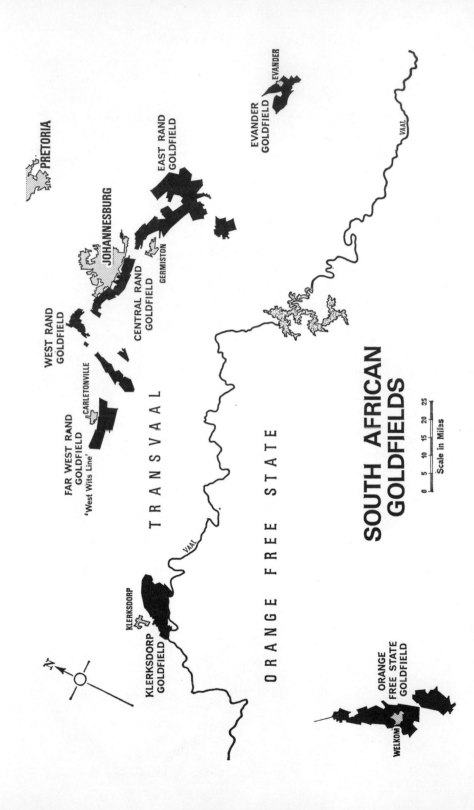

SOUTH AFRICAN GOLDFIELDS

3

South Africa

Going down a gold mine is rather like a trial run for Hades. You even leave all your clothes, including underwear, behind on the surface and, shrouded in white overalls, enter a steel cage which plummets through a mile of rock in two minutes. There below is a noisy, hot, wet world lit by the dancing fireflies of the lamps on miners' helmets. A ten-minute walk along a gallery cut through rock whose natural temperature is over 100° Fahrenheit, and any visitor is soaked by a combination of sweat and humidity. Then, above the constant hum of the air conditioning and the rumble of trucks along steel rails, comes the sound of compressed air drills biting into solid rock. On one side of the tunnel a narrow opening begins plunging down at an angle of nearly 25 degrees toward the bowels of the earth. It is barely 40 inches high and is delicately held open by props of blue gum. It is called, in mining parlance, a stope. Within the stope the rock seems to press in from all sides; tiny flakes fall from the roof into the pools of warm water in which everyone is kneeling or lying. Almost hidden in a fine spray of water to subdue dust, the long needle nose of a drill chatters into a hole in the rock marked with a blob of red paint. All along the side of the stope a continuous line of red paint highlights a 4-inch vein of rock that, even to the uneducated eye, looks markedly different from the rock above and below. It is a tightly packed bunch of white pebbles and between them, here and there, a minute speck of gold gleams in the beam of the miners' lamp. This vein or reef is the meat in the sandwich. This mine, Free State Geduld in the Orange Free State, is one of the very few in which the gold between the pebbles can actually be seen by the naked eye, for it is blessed with one of the richest reefs ever discovered in South Africa.

While most mines in South Africa boast only a quarter to half an ounce of gold per ton of ore, and it requires a magnification of at least two times to see even the largest specks of gold, Free State Geduld has an average of nearly one ounce per ton, making it one of the richest gold mines in the world today.

Extracting this thin seam of gold, lying at a depth of up to 2 miles or more below the surface, is an agonising and costly process. In human terms it costs over 500 lives a year in underground accidents in South Africa; in economic terms it has meant that the entire South African gold-mining industry has grown up with a structure and character very much of its own. Because the gold is so finely dispersed among those pebbless (a rock formation geologists call conglomerate), not only the narrow strip of gold-bearing reef must be blasted out and brought to the surface, much of the rock on either side of the reef must also be hauled up and crushed, for it becomes inextricably mixed with the conglomerate each time the stope is blasted forward by dynamite charges placed in the drill holes. In 1971, for example, nearly 80 million tons of rock had to be treated from South Africa's 44 mines, 'milled' to a fine powder and passed through tanks of cyanide solution to yield just 976 tons of gold.

That gold, however, was worth almost $1¼ billion and represented virtually 78 per cent of all gold mined that year in the non-communist world. South Africa's share is growing steadily. The industry there increased output every year except one (1967) between 1949 and 1970; it doubled during the decade from 1956 to 1966 and in 1970 set an all-time record of precisely 1,000 tons. Since the 1880s more than 30,000 tons of gold worth over $35 billion have been mined; in fact, some 40 per cent of *all* gold *ever* mined has come from South Africa. Thus she controls a unique position in the world of gold; a position which cannot be ignored by monetary authorities and may have won her special privileges from foreign governments who might otherwise have taken a tougher line against her racial policies. Certainly no new monetary agreement can be made without special consideration being taken of the South African case. And her gold is so abundant that it inevitably dominates the free market price. When South Africa held back a good third of her production in 1972, the gold price on the free market went soaring through barriers that many experts had not expected to see broached for a decade. She is, in fact, gold supplier to the world. Her only potential rival, the Soviet Union, whose gold output is estimated at just over 200 tons a year, trails far behind.

At home gold has been closely interwoven for almost a century with South Africa's economic and industrial development. As early as 1910, gold made up 60 per cent of her exports. The Dominions Royal Commission of 1914 estimated that 45 per cent of the Union's total revenue was attributable directly or indirectly to the gold-mining industry. Even today, although strenuous efforts have been made since the 1930s to give South Africa a more broadly based economy, gold accounts for more than one third of South Africa's exports (compared with diamonds, which account for only 10 per cent) and it is the Republic's major foreign exchange earner. Within the country, the industry spends $600 million a year on supplies and equipment, quite apart from $430 million in salaries and wages to 410,000 people (43,000 whites and 367,000 Africans). Gold, say the South Africans, has been the fly-wheel of their expansion.

The gold comes from the 300-mile-long arc of gold fields spreading out to the east and west of Johannesburg in the Transvaal and sweeping down into the Orange Free State to the southwest. The original outcrops of the gold-bearing reefs, which were discovered on Langlaagte farm in 1886 (see page 40, have long since been mined out, but through some remarkable geological detective work they have been tracked far underground to the Far West Rand gold field 40 miles west of Johannesburg, to Klerksdorp another 50 miles farther west and on to the mammoth Orange Free State field 200 miles away. There is even the maverick gold field of Evander, 80 miles to the southeast of Johannesburg, which is outside the main Witwatersrand basin that embraces the other fields. Unlike other gold rushes, where miners flocked in for a few years, panned the cream of the alluvial gold and went on their way, South Africa has had a permanent gold rush for the last ninety years.

As fast as one field was exhausted, another was found. In 1949 the Central, East and West Rand gold fields in the immediate vicinity of Johannesburg produced nearly 90 per cent of South Africa's gold. In 1971 they contributed barely 10 per cent, while the Orange Free State field, which did not come into production until 1951, yielded 35 per cent of the total output. The other new fields of the Far West Rand, Klerksdorp and Evander contributed 32 per cent, 16 per cent and 7 per cent respectively. Thirty-three new mines have been brought into production since 1945.

Right from the start, South African gold mining was not for the individual miner. The fact that the gold-bearing reefs dipped underground at an average angle of 25 degrees after outcropping near Johannesburg, meant that large companies able to muster both capital and expertise were essential for the deep mines. Even in the early 1890s it cost up to $2 million to start a new mine. By the 1960s a mine like Western Deep Levels, which will eventually burrow down to a depth of 16,500 feet, cost $95 million to bring into production. Thus, while the industry was still in its infancy, the great mining finance houses, Rand Mines, Gold Fields, Johannesburg Consolidated Investment, General Mining and Finance Corporation and the Union Corporation were established. Only two newcomers have succeeded in gate-crashing this hard core of five since the 1890s. First came Sir Ernest Oppenheimer with Anglo American in 1917, then A. S. Hersov and S. G. (Slip) Menell founded Anglo-Transvaal Consolidated Investment (now Anglo Vaal) in 1933. Together these seven companies have shaped and stimulated the world's leading gold-mining industry. Although each individual mine is floated as a separate company, with its own shareholders, chairman, board of directors and mine manager, its destiny is guided by one of these giant seven. They provide the financial security, administrative experience and technical know-how that a mine could not hope to develop alone. Each house also operates a central buying service for the mines under its wing. A nominal commission is charged, but the saving through bulk ordering is much greater. The mine manager may be a king—or a tyrant—within the confines of his own mine, but nothing humbles him faster than a summons to call on the finance house's consulting engineer at head office in Johannesburg.

The comforting knowledge that a finance house is administering a new mine is crucial in persuading an investor to risk his money in the venture. Geologists, after all, have been known to make mistakes, and the initial vast investment in a mine is staked on the evidence of a handful of cores from boreholes, which in themselves have not produced enough to pay for a pack of cigarettes. Before World War II thousands of small shareholders, particularly in Britain, provided the industry with much of its capital.

But in the postwar decades, rising death duties plus controls on the export of capital from Britain have curtailed this channel of investment. The sheer increase in the cost of establishing a mine has also made it impossible for the ordinary shareholder to bear the brunt of the burden. New finance, therefore, has come from insurance companies and similar 'institutional' investors. But because these investors must have a guaranteed return, the capital has been advanced only to new mines backed by mining houses of the highest repute. Even then it has frequently been in the form of loans, which can later be converted into shares if a new mine is, in every sense, a gold mine.

The leading finance house is Anglo American, whose solid looking headquarters at 44 Main Street tower over Johannesburg's business centre. When it was built in 1937, Sir Ernest Oppenheimer told his architect, 'I want something between a bank and a cathedral.' From his high altar on the first floor Sir Ernest dominated the postwar pattern of South Africa's gold-mining (and diamond) industry. The decisions that he made influenced the character of the industry long after his death in 1957.

The Anglo empire now administers eleven gold mines, producing about 40 per cent of South Africa's gold, and has a stake in sixteen other mines (quite apart from their diamond off-shoot, De Beers). Their main strength is in the Orange Free State gold field, which has mushroomed out of the open veld since 1946. Here their seven mines provide two-thirds of the group's gold output. Yet, when Sir Ernest Oppenheimer first took the bold decision to go ahead with development of the field, there were no roads, railways, power or water supplies into the area; the few farms were reached over unpaved roads. Twenty years later there is the flourishing new town of Welkom, with a population of 132,000, complete with schools, churches, banks, hotels and a swimming pool built to Olympic standards.

Sir Ernest was the last of a great line of South African mining entrepreneurs like Cecil Rhodes, Barney Barnato and Guy Carleton Jones, who had the flair and courage to stake their reputations on the potential of a new gold field. 'The Orange Free State Gold Field was a gamble,' said Adriaan Louw, chairman of rival Gold Fields, which also took a risk in the area and failed. 'No one would start a mine today on the basis of the information that was available on Free State in 1946.' But because Sir Ernest Oppenheimer took the plunge for Anglo American, the South African industry won a new lease on life. The new field, he said, 'is the most significant in South Africa since the finding of diamonds at Kimberley and gold on the Witwatersrand.' Three of the six 'super' mines, producing over 40 tons of gold each a year, are in the Orange Free State; the mines, Free State Geduld, President Brand and Western Holdings, yielded 155 tons of gold between them in 1971. Although the Orange Free State is predominantly Anglo territory, Anglo-Vaal, Rand Mines and Union Corporation each have one mine there.

Sir Ernest Oppenheimer's last gamble before his death gave Anglo yet another of the present 'super' six mines— Western Deep Levels in the Far West Rand field. The debate on whether or not to start Western Deep raged from 1943 to 1957, for the proposed mine presented major financial and technical hurdles. It would burrow down initially to 13,500 feet, where a rock temperature of 130° Fahrenheit was anticipated. Just to complicate matters further, the mine lay in the Gatstrand Hills which contain vast quantities of underground water; each shaft would need pumping capacity to cope with 30 million gallons of water a day. The promise of reward, however, seemed to Sir Ernest Oppenheimer, at least, to justify the development costs of $95 million. The best encouragement came from two mines that had already been sunk to the north of Western Deep

Rand Mines' Blyvooruitzicht and Gold Fields' West Driefontein. Both had revealed remarkably high grade ore, and West Driefontein was already moving into first place as South Africa's largest mine. Boreholes had shown that the rich Carbon Leader Reef, from which these two mines got much of their gold, plunged beneath the proposed mine property at a depth of between 2 and $2\frac{1}{2}$ miles. There was the added bonus of another gold-bearing reef, the Ventersdorp Contact Reef, much nearer the surface. It was this that made Western Deep a practical proposition. Initially the Ventersdorp Reef could be mined to produce revenue, while the much deeper shafts plunged on toward the Carbon Leader far below. Thus, in July 1957, Sir Ernest watched the first shaft-sinking drills commence operation on South Africa's most ambitious mine. It was a final fling that paid off handsomely. Western Deep Levels came into production in 1962 and is now regularly producing over 50 tons of gold a year. Over an estimated life of sixty years the mine is expected to yield 2,000 tons of gold.

Under Sir Ernest's son, Harry Oppenheimer, Anglo's enthusiasm for gold seems more subdued. No new gold field has been tracked down since 1950, while rising costs, without a compensating rise in the price of gold until 1968, have diverted their energies into other spheres. Their only major expansion in gold has been the South Vaal extension to their Vaal Reefs mine in the Klerksdorp field. 'You can't sit with your money waiting for a new gold field,' Harry Oppenheimer remarked to me in Johannesburg in 1967. Accordingly, although 28 per cent of the group's income was from gold in 1971, they have been busy diversifying out into everything from breweries to steel plants and paper making to property.

Anglo towers over the South African gold-mining industry, but two other mining finance houses have played crucial roles in determining its expansion since the war.

While Anglo pioneered the Free State field, Gold Fields of South Africa takes prime credit for the Far West Rand field, and the Union Corporation for the youngest gold field at Evander, which produced its first gold in December 1958.

Gold Fields of South Africa is the only one of the seven finance houses linked directly to a parent company outside the Republic. The London-based group Consolidated Gold Fields have a 49 per cent stake in Gold Fields of South Africa; Gold Fields of South Africa was actually a wholly owned subsidiary until 1971, when it was merged with West Witwatersrand Areas Limited, in which Consolidated Gold Fields also held a considerable position. Since 1960, however, Gold Fields of South Africa has had direct responsibility for the administration of all the group's companies in Africa, including their seven gold mines.

Gold Fields was born originally from the partnership between Cecil Rhodes and Charles Rudd in the 1880s, but its continued existence as a major force in the gold-mining industry is largely due to a Canadian-born mining engineer, Guy Carleton Jones. After being trained as an engineer at McGill University, Carleton Jones moved to South Africa in 1913 and joined Gold Fields—despite the advice of friends that mining prospects in North America were much better. His promotion within Gold Fields was rapid; he was manager of their Sub Nigel mine when he was thirty-three and in 1930 was made the company's consulting engineer. He found a group discouraged by the rising costs of the late 1920s, with little faith in the future of the Witwatersrand; the company was busy diversifying into other industrial interests around the world. Barely had Carleton Jones taken up his new position, however, when a young German geologist, Rudolf Krahmann, who had just emigrated to South Africa, called with the idea that Gold Fields should use a newly developed magnetometer to try to locate those elusive gold reefs. The magnetometer, he hastened to explain, would not locate the gold reefs themselves, but could pick up the pattern of magnetic shales of the Lower Witwatersrand system. Since the position of the gold reefs in relation to these shales was already known, they could then quickly be charted. Carleton Jones himself had never been satisfied with borehole results from an area about 80 miles west of Johannesburg, where surface formations suggested the reefs might be far below. He decided to give Krahmann and his magnetometer a three-month trial for a fee of $830, plus $560 expenses. It was the best and cheapest investment in prospecting ever made in South Africa. The magnetometer indeed revealed the pattern of the reefs. Carleton Jones then staked his whole professional reputation in persuading the board of Gold Fields, numbed by the depths of the Depression, to invest in this new gold field of the Far West Rand.

Today a huge three-dimensional model of the new gold field, which is most frequently dubbed 'the West Wits Line,' is proudly displayed in a special 'models' room on the ninth floor of Gold Fields' head office at Fox Street, Johannesburg. It reveals, even to the uninformed, how right Carleton Jones was to overcome all opposition in championing its cause. Not only did the Main Reef Series of gold-bearing conglomerate underlie the new field, but two hitherto unknown gold-bearing formations, the Ventersdorp Contact Reef and the rich Carbon Leader, were also buried deep in the field. The field was not developed as fast as the Orange Free State because the gold lay buried much deeper, but in the last twenty years it has advanced rapidly.

Today the Far West Rand area is the scene of greatest activity in South African gold mining. While the Orange Free State has really passed its peak, the 'West Wits Line' is still unveiling its true gold content. The pride of the gold field is West Driefontein, which started up in 1952 and seems to break records with whatever happens there. It is, quite simply, the most abundant mine the world has ever known. 'West Driefontein, is an outsize mine in every sense,' a Gold Fields manager said to me. 'It has the richest grade of ore now being mined in South Africa, it is the largest producer in the world—and it has the biggest disasters.'

West Driefontein is the one mine in South Africa that was still turning in an average grade of one ounce of gold per ton of rock in 1971; its output that year was almost 89 tons—more than the total output of all the mines either in Canada or the United States. And it has survived disasters that have frequently threatened to engulf it; deep-seated fires and torrential floods have all been overcome. For the last decade more than 30 million gallons of water *a day* have had to be pumped out of the mine. In October 1968 a major flood that sent a 100 million gallons of water a day surging through the mine and threatened to drown it for ever, was overcome through the sheer skill and deter-mination of mining engineers who managed to seal off the inflow with enormous concrete plugs. The mine seems to have an almost human resiliance to take such troubles in its stride, and simply goes on turning in high grade ore year after year.

West Driefontein's success has lead Gold Fields to invest over $80 million in developing another mine, East Driefontein, next door, which was formally opened in October 1972. The newcomer was expected to produce a good 20 tons of gold in its first year. Besides these mines Gold Fields administers four others, Doornfontein, Kloof, Libanon and Venterspost, in the Far West Rand. In all, the groups mines there yielded 177 tons of gold in 1971, which was 17 per cent of South Africa's output (Gold Fields only other mine is Vlakfontein in the old East Rand gold field). Just to make their position in the Far West Rand even more commanding they have another potential gold mine up their sleeve on a farm called Deelkraal just to the south of Doornfontein. Deelkraal was not a paying proposition as long as gold remained at $35 an ounce, but at $50 or more it could be a fine investment, and may be developed in the late 1970s.

The decision on whether or not to go ahead with Deelkraal underlines the hazards of the gold-mining business It is always a gamble to start any new mine on the basis of a handful of borehole results which have yielded a few little specks of gold, but at least with mines floated in the last thirty years the finance house could assume that the gold price would be steady at $35 an ounce. Therefore if the borehole results showed a good enough grade to pay at $35, they could take the plunge. The difficulty for Gold Fields over Deelkraal, however, is that borehole results indicated it will pay only at about $50 an uonce. And can one be sure in 1973 that the gold price in 1979 (which is the earliest time the new mine could come into production) will be $50? The trouble is they have to spend $100 million without knowing the answer for certain.

If Deelkraal is developed, it will certainly help the Far West Rand overtake the Orange Free State as South Africa's number one field by the end of the 1970s. Which shows how right Carleton Jones was back in the 1930s to spend $1,400 on Rudolf Krahmann and his magnetometer. Carleton Jones himself played a major part in the shaping of the West Wits Line right up to his death in 1948. He was appointed Gold Fields' resident director in South Africa in 1933 and in that post was responsible for all the early planning of Venterspost, Libanon and West Driefontein. His contribution to the gold-mining industry went far beyond his work for Gold Fields. The group's chairman, Robert Annan, paid this tribute to him in 1947 when he had to go into partial retirement through ill health: 'What is now proved (i.e., the existence of gold-bearing conglomerate)

was then only surmise and it required a rare degree of
courage, combined with technical and organising skill, to
produce evidence that would carry conviction and secure
financial backing for development. It is, I feel, the greatest
single contribution to the prosperity of South Africa's gold-
mining industry since the introduction of the cyanide
process in 1890, and deserves to be recognised as such.'
As a suitable memorial to the man whose persistence
established the West Wits Line, the new town that has
grown up to service the mines is called Carletonville.

Away to the east at Evander an even newer community
has mushroomed on the veld since 1956 to house the miners
working in the four gold mines in this small gold field
outside the main Witwatersrand basin. The mines are all
administered by the Union Corporation, which has also
borne the responsibility for creating the new town. The
nearest major supply of water was 50 miles away; main
sources of electricity were almost as remote. So, as the mines
went down, Evander went up under the watchful eyes of
the Corporation's estate manager, Bill Forsyth, who became
a combination mayor and father confessor to this infant
community of 32,000.

The real problem facing Forsyth, and the developers of
the other new mining towns of Welkom and Carletonville,
is to make them viable communities. 'A mine is a dying
source of employment from the moment it starts working,'
said Forsyth. 'It has only a limited life.' To build a permanent community which will not become a ghost town one
day, we must attract other industries. We need a community
that is not based simply on gold mining. At the moment
people come here to earn their bread and butter, then leave
when they retire. We have no old people.' So far it has
proved hard to persuade other industries to move in. At the
moment, in the first flush of prosperity from the gold mines,
this is no great worry, but in another twenty years there

could be weeds growing on the golf courses and basketball fields; the swimming pools might be empty and cracked. The problem is particularly acute at Evander because the young field has not proved such a good gamble as the Orange Free State or the Far West Rand. The four mines there (Bracken, Kinross, Leslie and Winkelhaak) produced only 62 tons of gold in 1971; Leslie, in particular, has been disappointing and may well pack up before the end of the 1970s, after a working life of less than twenty years. The trouble is that the gold reefs there have proved even more unpredictable than usual; extensive faulting in the rock over 5,000 feet below ground has meant that the reefs just are not where the geologists predicted. As a Union Corporation executive conceded sadly in 1972, 'I'm afraid we got our fingers burnt at Leslie.'

Anglo American, Gold Fields and Union Corporation administer between them 27 of South Africa's gold mines, producing 70 per cent of South Africa's gold. The other four mining houses share the rest. Rand Mines (which is actually closely linked to Anglo American) has five mines; Anglo Vaal and Johannesburg Consolidated Investment (in which Anglo American also has a considerable stake) and General Mining and Finance Corporation each have three.

However, although each mine is formally under the umbrella of one mining finance house, its board may well include directors from several others. There is an incredible interlocking of directorships and shareholdings both in the finance houses themselves and in the individual mines. Anglo American, for instance, has a significant investment not only in Rand Mines and Johannesburg Consolidated Investment, but also in Gold Fields of South Africa; and it holds stakes in sixteen mines that are administered by other finance houses, including Gold Fields' West Driefontein and Kloof and three of Union Corporation's mines at Evander. Gold Fields and Anglo Vaal, in turn, have directors on the boards of Anglo's Western Deep Levels, while Union Corporation is involved in Anglo's Western Holdings. This may seem akin to Ford having directors on the board of General Motors, but it arises because of the very special nature of the gold business. There is no place for competition between the mining houses as there would be if they were each trying to corner the market in cars, soapflakes or breakfast food. It is not even necessary for the controlling house to own more than half of the shares in a mine. Most houses administer with a 25 to 30 per cent holding, and will even run this down once the mine is well established. Anglo American held less than 10 per cent interest in its Springs Mine when it finally closed. In theory, the other shareholders could easily oust the mining house from control. It happened once, but it is not a manoeuvre to be undertaken lightly, for the new controller would have to provide the full administrative, financial and technical backing of a mining finance house.

The only real competition in the gold-mining business is at the prospecting level. Here the gloves are off as the geologists and geophysicists of the major houses try to outguess and outwit each other in detecting the most subtle clues to the location of a new gold field. Guy Carleton Jones exercised the utmost secrecy when he first sent Rudolf Krahmann out with his magnetometer.

There is always an air of considerable excitement in a consulting geologist's office on the morning that a potential gold-bearing core is brought in from drillings in the field. Every single inch of the core from a borehole 10,000 or 12,000 feet deep is carefully examined and always preserved. At the first sign of gold-bearing conglomerate, the consulting geologist at the head office in Johannesburg is immediately advised and the core is rushed to his office. If he agrees with the preliminary diagnosis made by his geologists in the field, he will probably show the core to his chairman or a senior director before it goes to be assayed. This is to protect his own interests, for there have been rare instances of geologists, with a weakness for playing the stock market, 'salting' the cores in co-operation with the assay office to get a much higher gold reading. One borehole result does not make a gold field, but a very high gold content in a single core could send speculators rushing for shares and push up the price. Most mines are floated only after a veritable patchwork of holes has been drilled over the proposed property. Gold Fields put down twenty-seven boreholes before deciding to go ahead with East Driefontein. But one highly profitable mine, Blyvooruitzicht, was floated by Rand Mines on the basis of a single borehole result.

A scramble for land may also develop if a mining house finds strong indications of gold, but does not own all the rights to the surrounding area which should logically be included in any new mine.

'The only time relations become strained,' admitted Michael O'Dowd of Anglo American, 'is when an area that ought to form one mine is fragmented between finance houses.' When Anglo wanted to commence the South Vaal extension of its Vaal Reefs Mine in the Klerksdorp area, it found that Union Corporation and Johannesburg Consolidated Investment both held some of the land. 'There was,' recalled O'Dowd, 'some very hard bargaining' before it was agreed that Anglo should administer the new mine. A similar struggle was taking place early in 1973 over Gold Fields proposed new Deelkraal Mine; Anglo American also held property alongside the potential mine area, part of which would inevitably have to be included in the new development. And Anglo were busy with their own boreholes. The issue turned on whose piece of land eventually yielded the best grade from the boreholes. Whoever proved to have the richest terrain will develop the mine and extend it into the other's territory. However, once the decision to float a mine has been taken, the need for cloak-and-dagger secrecy is over and the promoting house will offer shares to all the others.

Co-operation between the mining houses is enhanced by the Chamber of Mines of South Africa to which all belong, together with more than a hundred individual gold, uranium, diamond, platinum and copper mining companies, and collieries. The Chamber dates back to 1889 and throughout its history has helped to co-ordinate overall industry policy, research and the recruiting of labour. It is also responsible for the refining of all the gold at the Rand Refinery at Germiston, near Johannesburg. Each mine smelts its gold only into rough bars of about 1,000 ounces containing between 85 and 90 per cent pure gold. These bars are then delivered to the Rand Refinery, where they are refined up to 996 parts per thousand pure gold and cast into the standard 400-ounce 'good delivery' bars that circulate internationally. The refinery, the largest in the world puts through over 3 tons of gold each working day. The finished bars are then taken over by the Reserve Bank of South Africa, which handles the marketing of gold for the Treasury. Gold that is destined to become part of South Africa's official reserve is deposited either at the Reserve Bank's head office in Pretoria or at its branches in Cape Town and Durban. But gold for the free market is traditionally shifted direct from Germiston to Durban in a fleet of special trucks and then loaded onto Union Castle boats bound for Britain. Since 1968, however, some gold has been airfreighted directly from Johannesburg to Zürich. All payments for the gold are made by the Reserve Bank to the Chamber, which distributes the money accordingly.

The Chamber's seven-man Gold Producers' Committee, made up of one representative from each of the mining

houses, is the inner sanctum of the industry. Its deliber-
ations are as secret as those of a cabinet. In its early years
the Chamber was an extreme organ of British imperialism.
'It was,' Harry Oppenheimer once remarked, 'a highly
Empire-minded body.' But as the industry has faced a
tougher and tougher fight since the war to meet rising costs,
so the character of the Chamber has changed to meet new
conditions. It now has Afrikaner members, a very important
step forward in bridging the long-standing gulf between the
English and Afrikaners in South Africa. For generations the
Afrikaners, engrossed in politics, looked with great suspicion
on those 'foreigners' who had established the gold-mining
industry in Johannesburg. At first they expected the
industry to last only a few years, like all other gold rushes.
Even when it became clear that this was a permanent rush,
the reconciliation between English and Afrikaners was a
long time coming. Now that it has been achieved, gold is
indeed a South African industry. The first tangible result
of the reconciliation was a merger in 1963 between General
Mining and Finance Corporation and a new Afrikaner
mining group, Federale Mynbou.

Since the mid-1960s the Chamber has played a more
active role in encouraging world-wide discussion of the price
and marketing of gold. It embarked also on a rather belated,
but extensive, research programme spending $4 million a
year, not only to try to solve the increasing problems of
deep mining, but to try to make some technological
breakthrough to keep down costs. Until 1961 research had
been undertaken by the individual mining groups. 'It is a
matter of regret that it was never done before on a co-
operative basis,' Dr. W. S. Rapson, the Chamber's research
adviser told me. 'It would have been a terrific benefit.'
Trying to make up the leeway, Rapson has established
laboratories for mining research, physical sciences and
human sciences. 'The first real difficulty,' he recalled, 'was

to get research going on the underground problems of gold mining. Experienced mining men didn't like the idea of long-haired gentlemen with scientific degrees coming down to tell them what to do.'

The most essential problem posed by such mines as Western Deep Levels and East Driefontein is that of mining at depths greater than man has ever ventured into the earth before.

'The gold has never bottomed out,' one mining executive pointed out, 'but the great limitation has been the cost of refrigeration and ventilation at depth.' In the Transvaal the natural rock temperature increases on the average 1 degree (Fahrenheit) every 100 feet; in the Free State it rises a degree every 70 feet. This calls not only for superb ventilation systems, but also for refrigeration. At the ERPM mine on the East Rand, which goes to a depth of over 11,000 feet, the refrigeration plant has the capacity to make one ice cube for a drink for every inhabitant of London every day. At Western Deep Levels the refrigeration capacity installed is equivalent to an office air conditioning unit every 3 feet of working space. Such a plant is extremely expensive. Rapson and his scientists have now suggested other methods of regeneration cooling such as by spraying the walls of the mine and the freshly broken rock with water during the off-shift period. This cools the rock and in turn the air circulates through the mine, thus achieving an improvement in working conditions in the following shifts. The Chamber's scientists are also testing special jackets containing a series of water pockets that can be frozen. The workers don the frozen jackets, which keep them cool for a couple of hours, and then, as soon as the jackets thaw, swop them for freshly frozen ones from a deep freeze in the mine.

Another major hurdle at depth is the danger of rock failures and rock bursts. The pressure built up by 2 miles or more of rock stacked above the mine workings is such that, on occasion, rocks will virtually explode. The Chamber's scientists have been able to show that rocks follow specific laws of clasticity. Thus, the stresses and strains imposed by excavations and the release of energy as excavations are extended can be calculated exactly. By using computers, excavations can be mapped out and constructed with the minimum of rock failures.

The biggest challenge is yet to be overcome. The most significant breakthrough toward reducing mining costs would be to devise a method of extracting the narrow gold reef without having to hoist up with it several feet of useless rock from each side. A rich slice of the Carbon Leader Reef may be 4 inches thick and contain 10 ounces of gold in every ton of reef. But to get it out, a stope 40 inches high has to get blasted; as a result there is only 1 ounce of gold for every ton of rock to be crushed. The real test, therefore, has been to devise a rock cutter that can scour out initially just the thin gold-bearing reef—like pulling the meat out of a sandwich. The Chamber has been trying out just such a gadget in several mines since 1967. It is a drill that chips away the rich flakes of the reef. Production prototypes are already being evaluated in several mines, but so far the rock cutter has made no immediate change to the fortunes of the industry.

Actually the gold-mining industry has been very slow over the years to develop mechanisation underground, because it has so frequently been cheaper just to send in another ten Africans, instead of inventing a machine. Apart from a period immediately after the Boer War, when indentured Chinese labour was brought in from the Far East for five years, the gold-mining industry has been a sponge for African labour—Bantu, as the South African government prefers to call them these days—not only from the Republic but from all the states of central and southern Africa. Given this vast pool of cheap, unsophisticated labour, which has virtually no alternative form of employment, a system of migratory labour has grown up, under the control of the Chamber of Mines, which supplies mines with 320,000 Africans a year with almost 100 per cent turnover. The Chamber runs two recruiting organisations: the Mine Labour Organisation (NRC), which supplies labour from within the Republic and from Lesotho, Botswana, and Swaziland, and the Mine Labour Organisation (WENELA) which recruits elsewhere in Africa.

Although many mines prefer native South Africans, particularly those from the Ciskei and the Transkei, they are finding it increasingly hard to compete with the lure of other, newer industries in South Africa which are paying higher wages, and, unofficially at least, are giving Africans more scope for advancement to positions of responsibility. Thus two-thirds of the labour force for the mines is engaged outside South Africa. The Portuguese territory of Mozambique provides 80,000 annually, chiefly from the Shangaan tribe, while more than 60,000 come from Lesotho and 80,000 from Malawi.

At its local recruiting offices in each country, WENELA signs up the Africans provided they are eighteen years old, weigh at least 110 pounds and pass a preliminary medical examination. They are then transported free of charge to Johannesburg in WENELA's own launches, buses and aircraft or in chartered trains. At the reception centre on Eloff Street in Johannesburg, which can marshal up to 4,000 new arrivals a day, they are all X-rayed by a battery of six machines at a rate of 800 an hour, given a further medical check and are dispatched to the individual mines. Initially, they sign a contract for nine months or a year, although this can be extended for up to two years. All foreign Africans working in the mines must return to their homes within two years of arrival in South Africa. Once on the mine, provided they pass another medical with the mine medical officer, they normally spend three or four days at the mine's training school learning the simplest mining tasks and the rudiments of the special esperanto of the mines— *fanakalo* (meaning 'like this'). This language, which is a mixture of English, Afrikaans and African dialects, is essential because the Africans in the mines are drawn from so many different tribes and speak fifty or more dialects. It is hardly a language designed for intellectual discussion, but it spans the needs of the mines. *Fikisa lo foshol*, for instance, means 'Bring the shovel'; *Nangu lo tshif-bas yena fika* translates as 'There comes the shift boss'; *Wena lova* is just another way of saying, 'You are loafing'. Gold itself is *golidi*.

The Africans also learn how to use a shovel, load a truck or haul the ore out of a stope that has been blasted. Aptitude tests sort out those with previous experience, who can graduate to more responsible jobs of driving underground trains, drilling or acting as boss boy for a gang. But the majority, who have probably come direct from a very primitive tribal society, begin with simple tasks. While the mines try hard to create an image of benevolence toward the Africans, they are held in contempt by many individual white miners. 'They are only just out of the trees and the tail has been nipped off,' said one senior instructor at the training school at Free State Geduld, as I went round. Once the Africans go underground they are paid a minimum wage of 50 cents (U.S. $0·65) for an eight-hour shift. This rises slightly after a month's experience. The more experienced Africans working machines and acting as boss boys earn considerably more—even up to $150 a month. But the average cash wage paid by the industry as a whole to its Africans in late 1972 was only 83 cents ($1·10) per shift. On top of this the Africa receives free food and accommodation plus free transport from his home to the mine and back at the end of his contract period. Even so the basic wage is so low, compared with U.S. or European standards, that it places the South African mines on an entirely different economic level from gold mines elsewhere. By comparison the lowest pay for underground workers with no previous experience at gold mines in North America is now over $150 a week. The real significance of this cheap labour is best revealed by the fact that the white South African miners earn up to twenty times as much as the black. The total annual wage bill for the 40,000 white mine workers is usually twice the amount for the 360,000 black. Furthermore the level of African wages has increased so slowly that, according to a recent major study, real earnings in 1969 were actually slightly *less* than in 1911.[1]

And the African has no real opportunity for advancement. Long-standing legislation, going back before the formal establishment of apartheid, does not permit the African to graduate to any skilled work in the mine. Nor is he permitted to form a trade union. In recent years there has been some attempt to work out a pension scheme for African miners, but this is complicated by the fact that the Africans frequently have no idea how old they are and also may change their names several times during their life. There is, however, a non-contributory system of ex gratia payments after fifteen years' service. A boss boy gets $40 for every year he has served, an ordinary worker gets $33 for each year.

While they are working in the mine, the Africans have free accommodation in a compound close to the shafthead. They sleep ten to twenty in a room, in bunks ranged around two tiers in the wall. On the new mines they eat free in a cafeteria-style dining room which is open virtually twenty-four hours a day. There is no limit to how much they eat, although meat is rationed on some mines (2

[1] Francis Wilson, *Labour in the South African gold mines*, Cambridge University Press, 1972.

pounds of meat, 1 pound of fish a week). In most compounds there is a bar which sells everything from whisky and vodka to kaffir beer made from fermented maize. The kaffir beer, containing 3·5 per cent alcohol, looks like cold cocoa. It is slightly rough on the tongue and tastes sour. It is hailed among white miners as an excellent cure for ulcers. There are film shows in the evenings and at weekends. Westerns are the favourites, but they are carefully censored beforehand. All ambushes of stagecoaches are scissored out to avoid Africans seeing violence and getting ideas of doing a little dry-gulching of gold themselves.

The boss boys normally share a room between four, while the native *indunas*, who work as liaison men between the white compound manager and the Africans, frequently have their own self-contained flats with living room, bedroom and bathroom. Married quarters, however, are virtually non-existent and are only normally available for Africans who work at the mine on a permanent basis as clerks. Many, in fact, are not married, and one of their main reasons for coming to work in the mines initially is to earn the necessary price to buy a wife when they return home. To help them save, the mines encourage them to have at least part of their pay deferred until they complete their contract or remitted direct to their families at home. For those from the tribal reserves in South Africa, it is up to each individual whether any of his pay is deferred; for those from outside the Republic, normally half their pay is deferred after the first six months of their contract.

Medical care is free throughout the contract period. There are thirty-five hospitals with eight thousand beds for the Africans working in the mines, most of them very well equipped. The Ernest Oppenheimer Hospital, for example, maintained by Anglo American at Welkom to serve the 50,000 Africans working in the Orange Free State gold field has 887 beds, two general operating theatres, plus three specialist ones for ear, nose and throat, orthopaedic and resuscitation work. The theatre block is air conditioned and all impurities are removed from the air by thermal precipitation. Medical gases are piped direct to all the operating theatres from cylinders in a special gas room. There are physiotherapy and occupational therapy departments plus an artificial kidney machine.

Well-equipped hospitals are the least the gold-mining companies can provide in this most hazardous of occupations. The mines try hard to instill a constant awareness of the need to observe safety precautions, but it is not easy to drill safety codes into primitive tribesmen who may have a fatalistic approach toward injury and death. At some mines everyone is schooled to say *Mina sindile* 'I am safe' as a standard form of greeting instead of 'hullo.' This emphasis on safety has resulted in the South African mines having a lower death rate than, for example, Canadian or American mines. But despite all the coaching, the death rate per thousand for Africans working at the mines is almost double that for Europeans.

The attitude to the African workes in the mines and the whole concept of cheap migratory labour are, of course, the most sensitive points of the South African gold-mining industry. The gold companies argue that without the abundant supply of cheap labour (and there are always more Africans in countries outside South Africa applying to WENELA to work in the mines than can be accepted), most of the gold mines would have been priced out of business years ago. And if they were put out of business, the Africans would have nowhere else to turn for employment. The theory goes that it is the lesser of two evils to employ 360,000 at a low wage, than, say, 150,000 at a much higher wage, with 210,000 then out of work. The $6 million that is remitted by the miners to their families in Malawi each year, and the $15 million to Mozambique, make an important contribution to the economy of those countries. A step further in the gold companies' argument is that most of the Africans have small farms or cattle at home and that their wages at the mines are purely a nice bonus. 'He comes to get pin money,' J. A. Gemmill, the general manager of the NRC and WENELA, told me. 'We've never said that our wage is sufficient for him and his family to live on, but he comes here to get money for extra luxuries.'

However, this hardly justifies denying the African any rights to improve his position. No matter how capable and loyal he may prove, he cannot normally bring his family with him, he cannot advance to a more responsible job, he is not allowed a trade union to plead on his behalf. The hopeful sign, however, is that the more conscientious executives are becoming alert to the basic indignities of the system of migratory labour. Harry Oppenheimer, whose Anglo American has long tried to pay better wages and improve conditions, told me in 1967, 'we have got to use our African labour much better. In particular we would like to use a much larger proportion of settled married labour.' But the keynote for the 1970s was really made by Bill Wilson, a deputy chairman of Anglo American in a remarkably frank (for South Africa) speech in Pretoria in September 1972. His remarks are worth reporting in some detail. He began by stating that 'our assessment of the labour situation is that our attitudes and practices towards black employees require major overhaul, . . . anyone involved in employment policies and practices in South Africa is constantly aware of the fact that double, treble or even quadruple standards are applied . . . what I believe is needed is that there should be one standard only and that it should be a high one.' African wages, he went on, were 'extremely low in an absolute sense and compare unfavourably with other industrial wages, themselves the subject of criticism.' Moreover 'there is a dangerously large and rationally unsupportable gap between annual earnings of whites and blacks . . . (and) in terms of job content more senior and experienced blacks are grossly underpaid as compared with whites.' Urgent action was required, he argued, to put this situation right. He suggested, for instance 'that in all our enterprises scientific job evaluations should be undertaken, reasonable and logical wage curves drawn and fair wage targets established, which are not related to the race of the incumbent.' At the same time Mr. Wilson called for improvements in 'the whole area of negotiation, consultation, grievance procedures, communication in general, all of which are at a low level in South Africa and which industry in its own interests must tackle.'

The question now is whether such good intentions can actually be translated into firm action. The problem is how far the South African government—and the white miners' union—are prepared to relent on such issues as permitting more of the Africans to live on the mines in permanent married quarters and to carry out more skilled and responsible jobs.

The industry took a preliminary step forward as far back as 1964 with a trial run at twelve mines under which African labour underground was given more responsibility. Previously, after blasting operations a white miner had to examine all working areas, including those in which blasting had not actually taken place, before the African gangs could move in to continue their work. Under the trial scheme, the African boss boy, instead of the white miner, was permitted to double check all working areas where blasting had not actually taken place. This aided productivity, since it saved the Africans' waiting for clearance and may even have improved safety. 'The experiments showed,' the Chamber of Mines President, Herman Koch, was able to report in his 1966 presidential address, 'that higher productivity could be achieved, enabling white miners and their Bantu assistants to enjoy higher earnings and improved status, and at the same time offering mines, especially older ones, the prospect of reduced working costs and longer lives.' What looked like a very significant advance in the history of the industry soon foundered, however, on reefs laid by an extreme right-wing element of the white Mine Workers' Union, who feared that giving the Africans more responsibility would take work from their own members. After opposition from the union, a government industrial tribunal ordered the suspension of the experiment.

Fortunately that was not the end of it. In the spring of 1967, after months of argument between the Federation of Mining Unions and the Chamber, a 'productivity bargain' was struck. In return for increased pay on a monthly salary basis, the white unions agreed both to the abolition of certain restrictive practices relating to the manning of new equipment and to African's holding positions of greater responsibility. In particular, they decided to allow the previous strict personal supervision of every move of African labour by white miners to be reduced to a more practical level. 'It's inevitable that the Bantu will get more responsibility underground,' said Adriaan Louw, chairman of Gold Fields, 'and if you give him more responsibility he'll thrive.' The powerful Mine Workers' Union, however, remains utterly opposed to African advancement, and particularly to the African being given a 'blasting ticket', which would entitle him to handle explosives. And unless that hurdle is overcome the African cannot take a great step forward.

However, there now seems little doubt that although the migratory labour system will persist in the foreseeable future, the African will be paid substantially more in the 1970s. The climate of opinion towards higher pay seems to be swinging fast. As a Chamber of Mines executive conceded to me in the autumn of 1972, 'We are going to have to hike wages fast, even if we don't get it back in productivity.'[1]

This is quite an admission, for 'productivity' has become the watchword of the industry these last few years and every advance so far has been tied to its improvement. Until 1968, when the introduction of the two-tier market for gold finally gave hope of a higher price, they had been fenced in by rising costs set against a fixed gold price. Costs per ton of ore milled more than doubled between 1949 and 1966. When I first met mining executives in Johannesburg early in 1967 they were very glum about the future of the industry. The mines were being forced to extract more and more of their high grade ore, and were leaving vast areas of low grade ore unmined. The Chamber of Mines had estimated that gold worth $1·4 billion was rendered uneconomic to mine in the five years 1961 to 1965. 'The Free State is being murdered, because it is producing too quickly,' Dr. W. J. Busschau, the former chairman of Gold Fields remarked to me then.

The industry's pleas for more help finally won a firm response in the spring of 1968, when the South African Finance Minister Dr. Diederichs announced a state aid scheme which was worth $114·8 million over eight years to ailing gold mines. The plan was aimed at enabling marginal mines threatened with closure to hang on for a few more years pending a rise in the gold price. To qualify for assistance the mine had to have a future working life of eight years or less, but had to contain significant ore reserves. Tom Reekie, then president of the Chamber of Mines, welcomed the news. 'The industry,' he said, 'believes that this scheme will help to tide over the assisted mines until there is a general increase in the price of gold.'

The assistance certainly provided a lifeline for many

[1] Anglo-American set the pace by raising its African mine employees' wages by an average of 26% from April 1 1973.

mines. In 1971 some twenty of the 44 gold mines were on
state aid, and received between them over $20 million. But
even then their future did not look particularly rosy.
Although the gold price itself has finally gone up substantially, it is still not really enough to make some old mines
more viable propositions. They need a steady price of over
$70 an ounce to make it really worthwhile going after much
of their low grade ore. And although the London market
price for gold was hovering around $65 for much of the
latter half of 1972, the mines themselves were not getting
the full benefit of that increase because South Africa was
selling only two-thirds of production. The mines, therefore,
were receiving only about $55 an ounce overall.

Still, the whole industry was much more cheerful by
1972. 'There has been a tremendous psychological change in
Johannesburg,' admitted one of Anglo American's directors,
'people are doing exercises to see how all the old mines may
be revitalised.' The real tonic was the price spiralling up to
$90 in early 1973, with the market easily absorbing all
South African production.

There is most optimism about the Orange Free State
fields, because it is relatively easy there to go back to areas
of low grade ore in producing mines, which had simply been
bypassed as uneconomic at $35. The old gold fields of the
East Rand, on the other hand, are a much tougher proposition. 'It is probably too late to go back to the mines there
if they are already abandoned and flooded,' the Anglo
American man pointed out, 'and even if the mine isn't
flooded the pumping burden on its old neighbours to keep
it dry may be intolerably expensive. The other problem on
the East Rand is that the reef keeps going down and down,
but the mines were never planned to follow them. East Rand
Proprietary Mine for instance, has a veritable rabbit warren
of shafts, but none can go deeper. If the mine had been
designed originally like Western Deep Levels then mining

could continue. So I suspect the deep old mines will grind to a halt despite the price rise.'

On the more prosperous mines, however, the immediate reaction to the higher prices prevailing after 1971 was to go for some lower grade ore. The result of this is that South African production fell slightly in 1971 and 1972 from its all-time record of 1,000 tons in 1970. On the other hand the income from gold was much greater (up $100 million in 1971) and the life of the industry has undoubtedly been lengthened. When I first visited the mines in 1967 there was an air of gloom everywhere that South Africa had virtually reached the peak of her production and that from 1970 onwards production would trail off very sharply. The Chamber of Mines had even calculated it would be under 200 tons in the mid-1980s. Five years later, however, there was a new air of confidence abroad. While the record of 1,000 tons in 1970 may never be broken, the output could settle down on a plateau of well over 800 tons annually for at least the next fifteen years. No other nation is going to unseat South Africa in the foreseeable future from her champion's seat in the world of gold.

4

Russia

Joseph Stalin and Bret Harte hardly seem compatible. Yet between them the Russian dictator and the chronicler of the American West gave Russian gold mining a powerful push into modern times and established the industry as second in the world, surpassed only by South Africa. Bret Harte, admittedly, played a rather passive role, but his influence was profound. During the late 1920s, when Joseph Stalin was moving into ascendancy in the Soviet Union, he became preoccupied with the growing power of Japan and that country's threat to Russia's Far Eastern possessions. There was little to tempt expansion of population or industry into the far wastes of Siberia, yet without stimulation of the economy of those regions, particularly the growth of urban centres and communications, Stalin feared that Japan might all too easily be able to establish a foothold. That was the moment when Stalin seems to have become fascinated with the California gold rush and its effect on opening up the American West. Stalin's passion for California was recalled years later by A. P. Serebrovsky, who was chief of the Russian oil industry in the 1920s until he was promoted by Stalin in 1927 to head a newly created Glavzoloto or Gold Trust. Serebrovsky's book *On the Gold Front* noted, 'Stalin showed an intimate acquaintance with the writings of Bret Harte. He said that the new districts of the United States were opened up from the beginning by gold and nothing else. "This process," said Stalin, "must be applied to our outlying regions of Russia. At the beginning we will mine gold, then gradually change over to the mining and working of other minerals, such as coal and iron."'

He had been greatly encouraged by one genuine gold rush in 1923–25 after a private prospector named Kuzmin found gold on the Aldan River, in eastern Siberia, south of Yakutsk. Overnight thousands of gold diggers from all the gold fields east of the Urals rushed to the tiny village of Nezametny, where the gold had been discovered. Within a couple of years over 12,000 miners, organised in co-operatives, had staked their claims along the river.

Stalin's eagerness for gold was purely practical and hardly fitted with communist theories on gold. Lenin, in an essay in 1921, *The Importance of Gold Now and After the Complete Victory of Socialism,* had stressed that while gold was unfortunately still a necessity in the capitalist present, it would become useless in the socialist future when the world was ruled by a communist society. Gold then, Lenin suggested, might be used only to cover the walls and floors of public lavatories. But until that socialist paradise was created, he conceded the Soviet Union must carefully save its gold, 'sell it at the highest price, buy goods with it at the lowest price. "When living among wolves, howl like the wolves." '

Stalin was clearly playing the pragmatic wolf when he called in Serebrovsky in 1927 and told him to start a major expansion programme for Soviet gold immediately. Adopting the same philosophy, Serebrovsky decided the best way to beat the wolves was to join them. Under the guise of being a professor from the Moscow School of Mines, Serebrovsky set off for the United States to study American gold mining. He included in his itinerary a thorough tour of Alaska, where the gold was mined in both geological and climatic conditions similar to the far eastern provinces of Russia. His guide for part of the Alaskan tour was a tall, powerfully built American mining engineer, John D. Littlepage. Serebrovsky was clearly impressed by his guide, for hardly had he returned to the Soviet Union, when Littlepage received the offer of a contract to go and direct the installation of modern machinery in the Russian gold mines and step up their production as fast as possible. Littlepage accepted and from 1928 until 1937, with only short spells of home leave, travelled from one end of the Soviet Union to the other by plane, train and Model A Ford. He supervised the creation of a fleet of ninety modern steam and electric dredges in the alluvial gold fields and powerhouses, mechanical hoists, crushers and cyanide plants in lode mines. A number of other U.S. mining engineers were hired on short-term contracts to work in selected mines across the Soviet Union. Since Stalin realised that the secret of California's success had been the opportunity for the ordinary man to achieve riches, individual prospecting was also encouraged. Men who found new gold fields were rewarded with gifts of up to 30,000 roubles. Special stores, much better supplied with food, clothes and luxury goods than any others in the Soviet Union, were established in the gold fields. They accepted payments only in gold or gold-backed certificates. The idea was not only to attract people to the region, but also to tempt the individual gold diggers to sell their gold in the stores, rather than salt it away. The authorities even hoped that apart from the newly found alluvial gold, the shops would also siphon out private hoards of gold held since before the Revolution.

Nothing on the scale of the Californian, Australian or Klondike gold rushes resulted from all this activity, but Soviet gold production surged ahead in the 1930s. Stalin's purges also provided plenty of cheap labour for the new gold fields; political prisoners were sent to the gold mines, as well as the salt mines, of Siberia. Indeed, until the mid-1950s, they were the prime source of labour there.

So Russia crept back up the world gold league table toward the top position from which she had been toppled almost a century before by the Californian gold rush of 1848 (see Chapter 2). By the mid-1930s production had more than doubled to 5 million ounces a year and she had overtaken the United States, Canada and Australia to move into second position in the league. Serebrovsky even boasted from time to time that South Africa would be overtaken by 1940. It was a boast that was not fulfilled. Serebrovsky himself vanished in the purges of 1937; his book on gold was hastily withdrawn from circulation. Since then a monumental silence has fallen over the whole question of Russian gold production and reserves. Even John Littlepage offered little guidance. He collaborated on a book *In Search of Soviet Gold* when he returned to the United States, but in it he carefully refrained from giving any production figures. He said he felt duty bound not to reveal confidential figures learned while in the pay of the Russians. He conceded only that Russia was comfortably ahead of the United States and Canada in production and repeated the boast that the Soviet Union had the potential to beat South African output.[1]

That target, however, has not been reached more than thirty years later. Although Russia's true gold output is still shrouded in mystery, it is clear that her production is far below that of South Africa. What is well established is that all gold production comes under a department of the Ministry of Non-Ferrous Metallurgy known as Glavzoloto, which also has responsibility for platinum and diamond mining. Glavzoloto, in turn, has two gold–platinum divisions; one for the Far East and Eastern Siberia (Glavvostokzoloto), the other for Western Siberia, the Urals and Kazakhstan (Glavzapadzoloto). Beneath them, the

[1] J. D. Littlepage and D. Bess, *In Search of Soviet Gold*, Harcourt Brace, New York, 1938.

day to day working of the gold fields is supervised by thirteen regional gold mining trusts.

Russia's richest gold fields are alluvial; indeed two-thirds of output comes from placer deposits, scoured out by over 100 big dredges. The most productive are in the extreme east of Siberia, in an area embracing the Lena and Aldan rivers, the province of Magadan and the Kamchatka and Chukotsk peninsulas, which jut out into the Bering Sea opposite Alaska. The gold fields, in fact, are found in strata very similar to Alaska and the Yukon.

Digging out the gold is no easy matter. The terrain is remote and harsh. The best deposits are often hundreds of miles from the nearest large towns or good roads and the harsh winters mean that work can often go forward for only five or six months each year. The chief problem is thawing out the permafrost that makes much of the ground where the gold lies permanently rock hard. Sometimes the top layers of ground are stripped and the site left open for two years to give the permafrost in the lower layers of soil a chance to thaw out. The Russians have even experimented with boring holes into the permafrost, filling them with water and heating it. At the world's most northerly gold field on the Karal'veyem River huge boulders are dug out of the permafrost and crushed to extract the gold.

Despite the hazards, these lonely gold fields in deepest Siberia yield a third of Russia's gold. The Severovostok regional gold trust at Magadan, which administers them, is expected to turn in at least 75 tons of gold a year. Apparently they do not always match up to this target. In mid-1971 the Severovostok trust was publicly criticised in the Soviet press for failing to reach the prescribed output. Some of the trust's managers were accused of 'irresponsible attitudes'. Their lethargy, the charges added, had created 'a worrying situation which jeopardises fulfilment of the annual Plan and pledges for extraction.'

The problem is that until the mid-1950s these fields had an abundant supply of forced labour. But now they have to rely on ordinary workers and they find it very hard to recruit. No one is clamouring to take a job in the depths of Siberia.

The managers were not the only ones in trouble for slackness. The individual gold diggers, who still roam the region panning for gold in small deposits that are not worth the gold trust exploiting mechanically, were also accused of not pulling their weight. These co-operative gangs, or *artels,* sell their gold to the trust's buying office, but appear to be fairly free to use their own initiative in where they pan. The criticisms seemed to stir them into action; by October 1971 they had already achieved their required output for the year.

The trust pays the *artels* about $43·40 an ounce for their gold (while the Soviet State Bank, in turn, buys from the trusts at about $65·40 an ounce). In the Severovostok trust area the *artels* contribute a significant amount of gold—probably up to 10 tons a year, almost 14 per cent of total output.

Compared to Severovostok, the yield of the other trusts is much smaller; the combined output of the other twelve regions barely equals Severovostok's 75 tons. The two most significant are the Yakut and Lena trusts, presiding over dredging and open-cast operations all along the Aldan, Yana, Lena, Vitim and Olekma rivers to the northeast of Lake Baikal in Siberia. The dredges, slowly gulping their way along these rivers, are the biggest in the world; one monster installed in 1969 on the Mamakan River (a tributary of the Lena) weighed no less than 10,300 tons, was almost 750 feet long and 130 feet high. A special road 90 miles long had to be built so that the behemoth could be conveyed to the river. Even so it can only work for a few months of the year; in winter the temperatures fall as low as −70 centigrade.

Relatively little Soviet gold comes from deep mines. There is nothing to compare, for instance, with the South African mines ferreting down after the gold reefs that plunge towards the bowels of the earth. The deepest Russian mine, Novaya in Kazakhstan, had only been sunk to a modest 2,000 feet in 1970.

But the Russians have made one unique development— the extracting of gold from sand on the seabed. The Severovostok trust is doing this with suction equipment in the estuary of the Oblukovina River on Kamchatka peninsula and the Primor trust is trying it out in the Bay of Sudzukh on the shores of the Sea of Japan. The actual yield of gold, however, is not great.

The best estimates suggest that the total gold production from all thirteen trusts working under Glavzoloto is about 160 tons a year. In addition a further 45–50 tons a year is produced as a by-product of copper, lead, zinc and nickel mining. So that overall more than 200 tons of newly mined gold is available in the Soviet Union annually.[1] This is about one-fifth of South African production, but three times the output in Canada, the world's third largest gold producer. So although the Russians have not yet come up to their own forecasts of overhauling South Africa, they are very firmly in the saddle as number two in the league. Moreover, the volume of their gold production does make them the only nation besides South Africa which can really influence the free market price of gold by substantial sales.

Looking ahead, their gold production seems set to increase steadily in the 1970s, especially with any encouragement from a permanent rise in the gold price. Two major conferences were called in 1968 to discuss the long-term expansion of the gold-mining industry. At the first con-

[1] These estimates are drawn chiefly from Michael Kaser's studies on Soviet gold, published in Consolidated Goldfields' reports *Gold 1971* and *Gold 1972*.

ference in Magadan, the director of Glavzoloto remarked that the industry was 'still in the development stage and no one knows what will be discovered at greater depths than at present mined.'

The higher free market price for gold prevailing in 1973 may well encourage them to undertake new developments. Their problem as long as gold remained at $35 an ounce was that their mining was quite uneconomic. Although there are no precise figures available on their mining costs, it is reasonable to suppose that they were at least as high as those of Canadian mines in similar terrain. The Canadian producers had a hard time keeping going with a government subsidy that rewarded them with nearly $50 an ounce. The Soviet State Bank's price of about $64–65 an ounce to the gold trusts, gives some indication of what costs must have been. But such expensive mining continued for the very good reason that gold has been the vital lifeline for obtaining foreign exchange both to buy abroad industrial equipment and essential wheat supplies after bad harvests.

The uncertainties of precisely how much gold Russia produces and at what cost have provided one of the best guessing games of the post-war years. The most lavish estimates at one time suggested that production climbed steadily after World War II to a high of over 500 tons in 1957. More conservative estimates, including those of the U.S. Bureau of Mines, for many years pitched Soviet output consistently at about half that of South Africa, with a peak of around 350 tons annually in the later 1950s and early 1960s.

In 1963, however, there was an abrupt somersault. The United States Central Intelligence Agency (CIA) issued a report which revised downwards the estimates of Russian production right back to 1940. According to the CIA, Russian output throughout the 1940s and 50s never totalled more than 120 tons in any one year and was usually about

100 tons. This was a drastic downgrading of the Russian industry. Not everyone was ready to accept it. Samuel Montagu, the London bullion dealers, took issue with the figures in their *Annual Bullion Review*. 'In 1956 we had good reason to estimate that the sales by the USSR of $150 million worth of gold (about 140 tons) were probably about one-third of the Russian production that year and again, in 1957, the sales of $250 million worth of gold represented less than half the then current production.'[1] Montagu were confident of their figures, for they had collaborated closely with Moscow in many sales of Soviet gold.

And it was the very size of these sales over a long period that prompted many people to believe that Russian gold production was much higher than it actually was. Between 1946 and 1965 the Soviet Union sold over 3,750 tons of gold, worth more than $4·2 billion. To begin with the sales each year were fairly modest, but for nine years, from 1957 to 1965, they were substantially more each year than what is now accepted as production. In those nine years the Russians disposed of almost $3 billion in gold through the London, Zürich and Paris gold markets. In two peak years, 1963 and 1965, they sold a total of 976 tons of gold, worth $1·1 billion.

Now, if one accepts Samuel Montagu's estimates of production, these sales could all have been of newly mined gold. But if one takes the CIA figures on the other hand (and there is reason to believe that they came from a reliable source, none other than the Soviet defector Colonel Penkovsky), then the Russians were selling not just all their new gold, but a substantial proportion of their gold reserves too. Indeed, it becomes clear that by 1965 those reserves must have very nearly run out. Precisely how much was in them is uncertain; the Russians have been as coy

[1] Samuel Montagu & Co., Ltd. *Annual Bullion Review*, London, 1963.

about the status of their reserves for a generation as they have about gold production. They last declared them at $840 million in 1935.

They also keep quiet about a handsome bonus they received in the winter of 1936–37, when they accepted the bulk of Spain's gold reserves as security for payment for arms and aircraft for the Spanish Civil War. At least 500 tons of gold worth $560 million was secretly transferred from Madrid to Moscow. At a banquet in the Kremlin to celebrate the arrival, Stalin is reported to have said, 'The Spaniards will never see their gold again.' They never have.

Even counting in this nest-egg, it is obvious that with the high level of gold sales the treasury was being rapidly depleted. The CIA suggested in 1963 that Russian reserves were down to $2 billion; the further sales that year and in the succeeding two would have drained them further to about $1 billion. One estimate puts them at 977·3 tons at the end of 1965.[1]

Clearly this was a dangerously low level for a nation such as Russia, which is not a member of the International Monetary Fund and, at that time, was not able to secure very large amounts of credit in the West. Abruptly, therefore the gold sales stopped at the end of 1965, and, apart from small amounts in 1967 and 1971, did not resume in quantity until 1972. The intervening years (coinciding with a Soviet five-year plan) have apparently been used to rebuild the gold reserves to a respectable level. This reconstruction has obviously been helped by the fact that the Russians enjoyed better harvests in the late 1960s and so did not have to go shopping in Canada for so much wheat.

Thus from 1966 virtually all the newly mined gold went directly into reserves, except for relatively small amounts devoted to industrial use. And since production has been increasing steadily from about 150 tons in 1965 to com-

[1] Michael Kaser in Consolidated Gold Fields report: *Gold 1971*.

fortably over 200 tons by 1970, the reserves now look much healthier. At the beginning of 1972 they were estimated at just under 1,900 tons—or double the amount six years earlier.[1]

Given this much more solid golden platform on which to operate, the Russians could again begin to think about gold sales to the West. Moreover, the expansion called for by their new five-year plan for 1971–76 and poor harvests in 1971 and 1972 clearly required substantial sales of gold. Accordingly, the Russians sold around 200 tons through Zürich and London in 1972, and were expected to dispose of up to 300 tons in 1973.

Clearly the Russians will be a significant force in world gold markets in the 1970s. Unlike South Africa, their gold production still appears to be rising slowly. And all the indications are that they have considerable reserves, particularly in deep mines as yet undeveloped. Since they have put their reserves in good order, they can sell steadily from new production. They have the added advantage of a very small home demand for gold for jewellery. Indeed, up to 1969 no newly mined gold was allocated to jewellers in the Soviet Union; they had to make do with re-working old gold. In April 1969, however, the price of gold jewellery was raised suddenly by about 80 per cent (to the equivalent of $350 an ounce for pure gold) and more jewellery appeared in the shops. Some of it was from newly mined gold. Precisely how much gold is now allocated for jewellery is not known, although estimates by Consolidated Gold Fields in their annual gold survey, put it at 10 tons in 1969 and 14 tons for 1971. Above this another 30 tons or more would be needed for use in electronics and other industries. But that still leaves Russia with plenty to dispose of on the free market.

[1] *Ibid.*

Moreover her bars are always particularly welcome on the markets in London and Zürich because of their exceptional quality. The Russians refine all their gold electrolytically up to 999·9 parts per thousand pure gold, whereas the 'good delivery' bar accepted on world market is only required to be 995 pure. South African gold is usually a modest 996. So there is a distinct preference among industrial users of gold, who require extreme purity, for the Russian bars if they can get them. Otherwise they have to buy the conventional bars and then pay for them to be refined again up to the higher purity. And one sure way of knowing if the Russians are selling gold is to find out whether 999·9 good delivery bars are on the market; they may not actually bear the markings of Russia's two authorised refineries, the State Refinery in Moscow and the All Union Gold Factory, because European dealers often 'launder' the bars through a local refinery before selling them, but the presence of any 999·9 good delivery bars is a clue the Russians are about.

As for those gold-plated lavatories predicted by Lenin, no visitor to Moscow has yet reported seeing any. The Russians seem to be far too busy working up new gold production and negotiating in fine capitalist style with European bullion dealers to have got around to them.

5

Canada, The United States, South America, and Australia to Zaire

I CANADA

A sad simple letter was posted to me from Dawson City in the Yukon in the autumn of 1966. The stationery was headed 'The Yukon Consolidated Gold Corporation', and the letter said, 'We are in the process of closing down our operations permanently, and will no longer be actively engaged in the mining of gold by the end of this year.' It was the final curtain to the Yukon gold rush of 1897. The electric dredges that once scooped gold from the gravel of Bonanza Creek, Wounded Moose Creek and Last Chance Creek in this harsh, lonely terrain only 250 miles south of the Arctic Circle are silent. A handful of sourdoughs remain to pan for gold, and perhaps be rewarded with an income of a few thousand dollars a year. But, for the most part, the 500 residents of Dawson City look for their future to a new asbestos mine nearby or try wooing tourists with nostalgic shows and souvenirs of the heady days at the turn of the century.

As Dawson City's gold becomes little more than a memory, so the gold-mining industry throughout Canada totters in steady decline. 'Gold mining, once our brightest star, has become a poor step-child,' one mining executive in Toronto reflected. Even the high free market price of gold in 1973 did not cheer things up much. While it helped some mines pay better dividends it was really too late to be much of a tonic. So from a proud peak year in 1941, when nearly 5·5 million ounces (172 tons) were wrested from the ground to help Canada pay for urgently needed war supplies, production slowly diminished to just over 2 million ounces (62 tons) in 1972. Although Canada still ranks third in the world gold production league, she is contributing only one-thirteenth as much as South Africa or one-third as much as Russia. The decline has been accelerating for years. Many long-established mines have closed down, squeezed in the nutcracker of soaring costs and exhausted ore reserves. The record total of 122 mines in 1948, had shrunk to barely 30 by 1972. The Kerr-Addison Mine at Larder Lake in northern Ontario, which in 1960 produced more gold than any other in the Western Hemisphere, yielded only 140,000 ounces in 1972, compared with 600,000 a decade earlier. And the mine is unlikely to last through the late 1970s. The Hollinger Mine at Timmins, Ontario, once the largest gold mine until it was surpassed by Kerr-Addison, closed in March 1968 after being worked on a salvage basis for several years. A few small mines are being developed in the Red Lake area on the borders of Ontario and Manitoba (notably Robin Red Lake Mine), but their gold cannot compensate for the closure of the large mines.

The industry has, in fact, survived for over twenty years on the lifeline of a subsidy paid by the Canadian government under the Emergency Gold Mining Assistance Act (EGMA). Since EGMA—or cost-aid as the Canadians call it—began in 1948 more than $270 million has been paid to ailing mines. Virtually 80 per cent of Canada's gold was produced under cost-aid up to 1971. Only one major mine, Campbell Red Lake, with an output of around 200,000 ounces a year, was not subsidised, while the balance of production came as a by-product of copper mining and so did not qualify for assistance.

Moreover, the industry was not subsidised out of any

desire to maintain gold production, but to ease the plight of small communities like Dawson in the Yukon, Kirkland Lake and Timmins in Ontario and Rouyn, Malartic and Val d'Or in Quebec. They were born because of and nourished by gold, but faced the prospect of becoming ghost towns if the mines closed. Old miners who had spent most of their working lives in the gold mines were naturally reluctant to be plucked from their tiny, friendly communities in the wilds to go to work in copper or nickel mines elsewhere. Not only the miners, but the shopkeepers and bar owners, whose livelihoods depended on the miners' pay packets, fought a long rear-guard action to keep the mines open. So EGMA was maintained through the years; each time it came up for renewal the industry got into a panic that it would be discontinued, but the government always came through at the last moment with more money. Cost-aid regularly paid out around $15 million a year. In 1966, when doomsday seemed near, a delegation from thirty-six mining villages petitioned the government in Ottawa, pleading for continued aid. They argued: 'The central issue is a humanitarian and defensive one, to preserve and if possible expand the local economic basis of the gold towns. We believe that the Federal Government should not lose sight of this issue, nor confuse the development and employment of people with the preservation and operation of municipalities. It may help individuals to retrain them and move them out to new opportunities. But it kills the towns they leave behind, loading them with economic burdens, ruining their standards of public service and creating as many problems as it solves.'

Their plea won them a reprieve both then and in 1970 when cost-aid seemed finished. Indeed, the subsidy has now been renewed through to 30 June 1976. A rather meaningless gesture as it turned out, because every mine came off cost-aid in the autumn of 1971 when the rising free market price of gold made it more profitable to forgo it. The subsidy was geared only to a continuing gold price of $35 an ounce (although this was adjusted to $38 in 1972). A sliding scale of payments was devised under which a mine became eligible for aid once its production costs rose to $26·50. The maximum payment of $10·27 an ounce was paid to a mine which had production costs of $41·58. That is to say it would have recieved $45·27 for its gold. Once the gold price went above that it obviously made sense to come off the subsidy. All mines receiving aid had to sell their gold through the Royal Canadian Mint. Until 1968 the Mint took the gold on behalf of the Canadian Exchange Fund Account, but after the Washington Agreement of March 1968 it sold the gold on the free market.

While cost-aid undoubtedly helped mining communities stay alive, it did not produce any long-range solution to the basic problem that known gold deposits, which could be mined at an economic rate, were rapidly running out. And since it could not be used for long-term exploration, it merely kept things ticking over, but did not pay for expansion. Consequently the mining companies became disenchanted with gold and directed all their energies and cash reserves into branching out into more profitable copper or nickel mining and joining the search for oil. And whereas once upon a time, all Canada's mining engineers cut their teeth on gold, a whole generation of young mining graduates have paid prime attention to copper, iron and nickel and are almost ignorant of the subtleties of mining for gold. The young unskilled miner has also been lured by the bulging pay packets of the base metal mines, where he can earn up to 50 per cent more an hour than in gold, in far less trying underground conditions.

So that while on the one hand old gold townships have pleaded for subsidy, the few new mines that have been developed elsewhere have found it extraordinarily hard to woo labour. The trouble is that most of the promising gold deposits are in the remote far north, a land of muskeg, with black flies in summer and blizzards in winter. No one is clamouring to leave the cities and head out into the muskeg to dig for gold. When the International Nickel Company stumbled on a sizable gold deposit at Contwoyto Lake in the Northwest Territories in 1963, they decided it was just not worth sinking a mine so far from civilisation. A complete new town would have been required to house the miners, who would have demanded, quite rightly, high wages and long holidays. Even more complex, all machinery and supplies would have had to be brought in by air; there were no roads for hundreds of miles.

The only mine that has made a go of it in the north is Giant Yellowknife on the northern shore of the Great Slave Lake, which is Canada's largest mine, with an output of 217,000 ounces in 1971. The mine is more than 700 miles north of Edmonton and is definitely for those who like the great outdoors. The mining company has established schools, a hospital, cinema, ice hockey and curling rinks and even a golf course to tempt miners and their families to this lonely nook of the world. For those who can brave the long winter, with temperatures hovering around minus 60° Farenheit, the summer brings rewards of boating and swimming in the Great Slave Lake and fine fishing for trout, bluefish and arctic grayling in the maze of nearby rivers. But such inducements rarely encourage men to stay long at the mine.

Some mining executives, however, have never let the depressing outlook for gold get them down. Their enthusiasm has survived. 'There is a satisfaction in finding gold that you don't get with other metals,' said a geologist in Toronto. While Arthur W. White, the president of Dickenson Mines, has always been an unashamed gold advocate. 'I'm a gold bug,' he confessed to me one day, sitting behind a broad desk in his Toronto office, wearing a gold tiepin, gold cufflinks, gold wrist watch and lighting his cigarettes with a gold lighter. 'Gold has been good to me,' he explained. 'It's been good bread and butter, so I feel an obligation to the industry.' His optimism shows some signs of being rewarded. Dickenson Mines bought out the neighbouring Robin Red Lake Mine from Noranda and Dome Mines in 1969 and developed it directly from Dickenson workings. Early in 1972 they began to find substantial high grade ore deposits that surprised even Robin Red Lake's management. In one small section they were even getting out over $3\frac{1}{4}$ ounces of gold from every ton of ore—a bonanza to a gold miner, especially with the price at over $60 an ounce.

The high price of gold in 1971 and 1972 was obviously a fillip to many other Canadian mines. Once the price had gone past $45 an ounce, it paid them to come off cost-aid and sell their gold directly on the free market. By early 1972 every mine was off subsidy. The actual marketing is handled either by the Noranda Sales Corporation, which sells Kerr-Addison's output and the 400,000 ounces of by-product gold from Noranda's copper mines, or by the Bank of Nova Scotia acting for several other mines. Since Canada's own gold requirement for jewellery and industry is small, most of the gold goes either to Central and South America or, increasingly, to the United States.

Although there were more cheerful faces along King Street West in Toronto, where most mines have their head offices, the overall prospect for the industry was not radically changed. In the first place the mining executives could not commit themselves to any new development programme based on just a few months' good gold price. But even if gold stayed at $70 or more, many mining men feel that the price rise has come too late to save the industry. Old mines have been neglected; shafts and workings are full of water.

Only a stable price of well over $100 an ounce would really make it worth while pumping them out and opening them up. Even for mines still in operation the prospect was not much better. The trouble, according to Paul Kavanagh, vice-president of Kerr-Addison Mines, is that 'additional reserves are commonly in the walls of stopes already mined, and it is usually impossible to re-enter old stopes economically. So Kerr can increase its reserves only in those areas not yet mined.'

The mines which will get the best bonus from the price rise are naturally the young ones, which can now be worked more thoroughly. Camflo, for instance, a small mine at Malartic in Quebec, producing just under 100,000 ounces a year, has a much brighter future.

But it is unlikely that overall Canadian production will increase. The decline that has already gone on for a generation will simply be slowed down. 'To my mind, it is unlikely that production will increase,' Paul Kavanagh told me in 1972. 'At $70 a grade of ·15 ounce still only represents $10·50 a ton; you cannot bring in a new underground mine with that grade in Canada.' Furthermore, there seem to be relatively few signs of new gold deposits in Canada anyway. Only two areas, Red Lake in Ontario and Malartic in Quebec have promising rich deposits still to be exploited.

So although Canada will certainly retain her position as the world's number three producer for a while, with her production settling on a modest plateau of around 2 million ounces annually, the heady days of gold rushes will fade further into history every year.

II THE UNITED STATES

For the nation that ushered in the new age of gold with the California gold rush in 1848, the United States cuts a disappointing figure these days. Those prospectors who went stampeding out to California in search of a fortune would hardly recognise the place. In 1971 a mere 3,000 ounces of gold were scratched out of the ground there, mostly by amateur weekend prospectors, compared with 3 million ounces in the hectic year of 1853. Indeed, America's total gold production today falls far short of that 1853 Californian record—a modest 1,475,000 ounces in 1972. Although that was good enough to win the United States fourth place in the world gold league, she trailed so far behind South Africa as to be almost insignificant. Her output, in fact, could meet rather less than a quarter of her own requirements for jewellery and industry; the demand for gold in electronics alone just about matched domestic production.

The higher gold prices in 1972 and 1973 have been a happy windfall, but they will not stimulate the industry like the last major rise in 1934. That jump, from $20·67 to $35 an ounce, sent thousands of unemployed Americans hurrying west in search of gold and U.S. production rocketed in the late 1930s to an all-time record of almost 5 million ounces in 1940. Since then, however, the story has been very similar to that in Canada—steady decline. The American mines have not even enjoyed the benefit of subsidies which have cushioned the Canadians and most other gold producers. The American government, stoutly defending gold at $35 an ounce, would not give their ailing gold-mining industry a single extra cent. For over a decade now output has hovered between 1·4 million and 1·8 million ounces a year, and more and more old workings have shut down. The last large dredge scouring out gold in the rivers of California packed up in 1968, while up in Alaska only one major dredging operation, on the Hogatza River in the Yukon River region, was still going in 1972.

The American Bureau of Mines 1971 survey of the industry listed only four true gold mines and one placer operation left among the top twenty-five gold-producing companies in the United States; in the other twenty gold was simply recovered as a by-product of copper, or copper-lead-zinc ores. Moreover, 80 per cent of all the gold came from just three gold mines and one copper mine.

Amidst this gloomy picture, however, one mine has a remarkable record. The Homestake Mine at Lead, South Dakota has been going strong ever since 1877 and is certain to reach its century. Homestake has already yielded well over 1,000 tons of gold and is the largest producer in the Western Hemisphere, consistently turning out 550,000 to 600,000 ounces of gold annually. In a hundred years of mining, the Homestake men have burrowed out more than 200 miles of tunnels and shafts beneath the Black Hills of South Dakota. And just to show that they were not getting tired, the mining company launched a five-year $8 million deep-level development programme late in 1970. They plan to dig down to rich new gold-bearing area nearly 7,000 feet below ground; if the ore meets up to their expectations, the mine should be flourishing well into its second century.

The chairman of Homestake for many years, Dr. Donald H. McLaughlin, was the dean of American gold mining and a great advocate for a major increase in the price of gold. Indeed, the South African always regarded him as one of their few true allies in the gold revaluation lobby. Reviewing the soaring costs of extracting gold McLaughlin once lamented to a mining conference, 'It's really so shocking that when I read these figures I almost burst into tears and wonder why we're in business.' But despite his depression, Homestake looks like being American champion for many years to come.

It's nearest rival is not a gold mine at all, but an enormous open-cast copper mine run by the Kennecott Copper Corporation at Bingham Canyon, 30 miles southwest of Salt Lake City, Utah. Here the lonely desert road suddenly opens up to reveal a mammoth amphitheatre nearly 2,000 feet deep and 2 miles across. It is the world's largest man-made excavation and the men and machines who toil in it look like children's toys. This gargantuan hole in the ground yields nearly 300,000 tons of copper every year, but mingled with the copper are tiny specks of gold. Once segregated out they add up to a tidy 340,000 ounces of gold each year, which is a handsome dividend for Kennecott and helps them sell their copper at very competitive prices. Almost 40 per cent of America's gold now comes from this Bingham Canyon hole and other smaller mines where gold is also a profitable by-product of base metal workings. Seven copper mines in Arizona, for instance, turn in a bonus of almost 100,000 ounces of gold a year, while the Mayflower Mine at Wasatch, Utah produces over 60,000 ounces from a veritable sandwich ore containing gold, lead, copper and zinc.

A bright newcomer joined these old-established gold producers in 1965 as a result of some smart geological detective work by two geologists, John S. Livermore and Robert B. Fulton of Newmont Mining Corporation. Scouting around the Tuscarora Mountains of northeastern Nevada they came upon a fine new gold field over 6,500 feet up. But it was a gold field with a difference; old-time prospectors would have passed it by, only new technological skills made it viable. The largest specks of gold were only 0·0002 inch across, and had to be magnified 1,800 times before they could be photographed. The $1\frac{1}{2}$ mile long outcrop of rock that Livermore and Fulton located was, however, so rich in these tiny particles that it could yield one-third of an ounce of gold for every ton of rock—a minute proportion to the uninitiated, but a very profitable deposit to the gold miner. Moreover, the gold-bearing rock was all on the surface, and could be garnered by open-cast digging, not by costly mining. So in April 1965 at Carlin, Nevada,

Newmont opened the first major new gold mine in the United States for half a century. It fulfilled its rich promise by yielding 261,000 ounces of gold in the first year, and then settling down to turn in around 200,000 ounces annually. Other promising deposits, christened with such delightful names as Bootstrap, were found nearby, thus ensuring the new mine of good life well into the 1970s.

The success of Carlin prompted a spurt of interest in resurveying many other areas of the American West in the hope that new deposits that had not been revealed by traditional prospecting methods could be traced. The Department of the Interior encouraged the hunt by initiating a $10 million research project of its own, known by the formidable title of the 'Heavy Metals Program'. The research was undertaken by the Bureau of Mines and the Geological Survey between 1966 and 1970. Besides exploring new terrain, they went back over areas known to contain some gold deposits to see if they could be mined profitably by new techniques. At the same time the Department of the Interior through its Office of Minerals Exploration, was encouraging small groups of prospectors to go out and hunt for gold. They helped such picturesquely named outfits as the Mugwump Mining Company and the Original Sixteen to One Mine Inc. to get out in 'them thar hills' looking for gold. If anyone struck it rich they had to pay the government a 5 per cent royalty.

The best bonus to come out of this flurry of activity in the late 1960s was yet another brand new mine, at Cortez in Nevada. Like the Carlin Mine, Cortez, which started in 1969, is an open-cast operation extracting tiny specks of gold that are invisible to the naked eye. The mine—perhaps pit is a better description—yielded over 150,000 ounces of gold in its inaugural year and improved on this in 1970 to turn in 209,000 ounces. This put it slightly ahead of Carlin and made it the third largest producer in America. But its success may be short lived; the reserves of gold at Cortez are limited and the newcomer may bow out within a few years.

The Heavy Metals Program also investigated improved methods of milling (i.e. crushing) and treating gold-bearing ores to see if it was possible to extract additional gold missed by traditional methods. As a result of their experiments, a new milling plant that treats carbonaceous gold ores chemically was opened at the Carlin in January 1971, enabling the mine to handle large stockpiles of ore that had previously been bypassed.

Whether this Heavy Metals Program will have any long-term effect on American gold mining is another matter. The Cortez mine certainly checked the decline in overall American production, but it did not make for a significant increase either. However, the five years of investigation do represent perhaps the most searching enquiry anywhere in the world made recently for new gold fields. Much of the work was done, of course, while the gold price was still at $35 an ounce. Clearly at that price there was little scope for expansion. But at $70 or $100 there may be more potential. The Bureau of Mines gold specialist, John West, observed in the autumn of 1972 (when the gold price hovered at $65) 'the feeling is that price levels of about $100 per ounce could really start a movement, but present levels are certainly not causing any stampedes.'

In the meantime much of the enthusiasm for prospecting comes from hundreds of amateur prospectors in California, who spend their weekends scrambling through the beds of the Trinity, Klamath, Feather, Yuba and American rivers that tumble down from the Sierras in northern California, hoping to find gold missed in the great gold rush over a century ago. Their equipment, however, is a little different from the old-time diggers; today's prospectors take along diving equipment to scout out underwater deposits beyond the reach of their ancestors. The basic equipment is a snorkel, face mask, weighted belt, fins, gold pan and crevicing tool (anything from a large spoon to a crowbar). A diving suit is not essential, but highly recommended in the chilly mountain streams fed by melting snows. A 'gold sniffer' made out of an old grease gun is also a handy extra. The real professionals, of course, take aqualungs and even float a suction pump along behind them on a little raft and try to suck up the gold with a gadget that looks like an outsize vacuum cleaner. The California Department of Mines and Geology helps them by publishing booklets on the art. 'Concentration of gold can be expected where the stream widens, near quiet pools below rapids, or along bends in the channel', wrote one geologist. 'Examine the stream bank for gold between the high water mark and the water level . . . Small roots, moss or other vegetable material near the water should be examined as they may trap fine particles.'[1]

Once in a while the ardant diver is lucky and ferrets out a 'virgin' crevice packed with coarse nuggets of gold. A niche on the North Fork of the Yuba River near Downieville yielded a handful of nuggets in 1958, one of which weighed 10 ounces.

The real problem, apparently, for any successful prospector is finding someone to refine his gold. Since 1968 the U.S. Mint no longer buys gold and the commercial refineries are geared entirely to the needs of major mines. The best bet is to find a friendly jeweller who will buy the nuggets if they are high quality. But any prospector who tries to remelt his own gold has to remember that Americans are not allowed to hold gold privately; only licensed commercial users can hold bullion.

Nowadays the mines themselves sell all their gold directly to the commercial market. Up to 1968 they could always

[1] William B. Clark, *California Geology*, published by Division of Mines and Geology of California Department of Conservation, June 1972.

sell to the U.S. Mint, which either kept the gold for America's reserve or re-sold it. Since that year the Mint has opted right out. The gold therefore goes directly to the free market, much of it through Engelhard Industries, the largest fabricators of precious metal in the United States. Engelhard, in fact, take so much gold, that their price is usually quoted as the American free market price. This means, of course, that the American mines have received the full benefit of the higher gold prices of recent years. While the extra cash has not yet given the industry a new lease of life, it has given it the first real bonus in thirty-five years. When I first talked to American mining companies in 1966 it was a question of how much longer they could hang on with gold sticking obstinately at $35 an ounce. Now at least everyone is holding their own. Homestake are busy expanding and rumours are trickling in of new explorations. Things are stirring. 'There is renewed gold exploration in the Leadville and Cripple Creek areas of Colorado and the North Moccasin Mountains of central Montana' says one report. With names like that to the fore, it sounds as if the old days might just be coming back.

III SOUTH AMERICA
The last real frontier of the world of gold is in South America, which, despite centuries of arduous mining by the Spaniards, still yields nearly 600,000 ounces of gold each year. Much of this gold comes from the narrow twisting rivers, first tapped by the Incas, that tumble down from the Andes in Colombia and Bolivia. Now, in place of the Incas, a horde of local gold diggers or *mazamoreros* scour the river like leeches, panning for gold. Once in a while they also partake in a little hijacking from the subsidiaries of the American group, International Mining Corporation, which controls the more formal gold production in both countries. International Mining, which operates the largest fleet of

gold dredges outside the Soviet Union (they have ten in Colombia and one in Bolivia), produces three-quarters of Colombia's gold; the *mazamoreros* scoop up the rest. It is an uneasy relationship, reminiscent of the old days in California or the Klondike.

'We have a bad problem with stealing,' complains Patrick O'Neill, the president of International Mining. 'They have laws in Colombia, but they don't seem to live up to them. We had one armed attack in December 1966, when some fellows on an island in the middle of the Nechi River opened fire on our launch that was carrying gold from the dredge. The launch driver promptly switched off the engine and jumped overboard, but the rest of our boys opened fire with shotguns and the fellows ran away. We had two men injured and wounded two of theirs.'

Things are not much easier down in Bolivia. The manager of the dredging operation was kidnapped twice in 1971 by local farmers, who were dissatisfied with lack of government action in building new roads in the area. The first time the manager was released for the ransom of an old tractor to be used in developing roads; the second time the deal was for a Land-Rover.

Even when there is no open violence, the *mazamoreros* give the mining company a hard time. 'When we get into shallow ground we get a hundred or more of them right in front of the dredge,' says O'Neill, 'and we have a hell of a time.' Aboard the monster dredges themselves some of the gold seems to vanish the moment it is recovered from the rivers. The manager finds it impossible to keep an eagle eye on the twenty-five men working aboard. Even if a man is caught stealing gold, legal actions often fizzle out. 'We catch some guys three times a week,' says O'Neill, 'but on the way to the local judge he throws the gold away. So there is no evidence and the judge turns him loose.'

Even in its underground mine at Frontino in Colombia,

which yields around 50,000 ounces a year, life is just as dangerous. Six men armed with rifles beseiged the gold-concentrating room in 1965. Sniping from the cover of the nearby hills, they wounded two security guards before the attack was repelled. One raider was shot dead at the mine, another died on the trail as he fled into the mountains. And it is as hazardous down in the mine.

'Outsiders called *machuqueros* get into the mine constantly,' Patrick O'Neill reports. 'They live for weeks at a time in the old stopes and our workers keep them supplied with food. They build small hand operated mills there and when they get enough gold, either take it out or send it out with our men. We make raids occasionally on these fellows, but usually catch them only when they are asleep, because there are so many old openings for them to hide in. It's dangerous for our people searching for them with a light—they may slug you or shoot you as you go by.'

Raiders are only half the battle in South America. Rivers can rise 30 feet in an hour after a tropical rainstorm, almost swamping the 140-foot long dredges as they wallow up the rivers devouring 6 million cubic yards of gravel a year. Mining camps are often miles from the nearest road and must be self-supporting. International Mining has established a 16,000 acre ranch at its Frontino mine with 5,000 head of cattle to feed the miners and their families. In Bolivia, where the company has a dredge operating in the Andes 80 miles northeast of La Paz, all equipment had to be flown in originally in two Northrop trimotor aircraft. Diesel fuel for the dredge was floated down a river in 1,000 gallon rubber tanks, until a road was eventually built.

Faced with such conditions, International Mining has been finding it increasingly hard to keep going. They are already shutting down their dredging operations in the Narino region of Colombia and moving the dredges to augment the fleet of their Pato and Choco operations.

But even with a higher price for gold these remaining operations may not continue indefinitely. The future of Choco and the Frontino mine is uncertain. The headache is that International Mining does not get the benefit of the new gold price in dollars. The Bank of the Republic in Colombia, to which they have to sell all their gold, pays a basic price of $38 an ounce, half in U.S. dollars and half in Colombian pesos. The company also gets 15 per cent extra in tax credits in pesos. Furthermore the Bank pays a bonus on 80 per cent of the production when it is eventually sold either direct to local jewellery users in Bogota or through the London and Toronto gold markets. This does give the miners about the London price (they were getting $65·26 in September 1972), but as most of it is in pesos, rather than dollars this barely covers supplies and American salaries. There is little incentive, therefore, to develop new production.

The other problem, of course, is that American mining companies are less and less welcome in South America. International Mining is already negotiating to sell a 51 per cent stake in their Colombian operations to local companies to satisfy the growing spirit of nationalism.

The prospect is that despite better prices for gold Colombian production will continue to fall; it has already dropped from almost 400,000 ounces in 1962 to 180,000 ounces in 1971.

Beside Colombia, the only other gold mines of substance are in Brazil. A group of mines in Minas Gerais, owned by Mineracao Morro Velho, have the distinction of being the world's oldest gold mines still in production. The mines started in 1835 and were still doing well in 1972, turning out 155,000 ounces each year. They sell all their gold directly on the free market in Brazil and have benefited considerably from the higher prices of recent years. In fact the old mines are as hearty as ever. They embarked on a major expansion programme in 1971 aimed at pushing their production up by over 50 per cent in three years. Even its proved reserves (and miners are always cautious on defining how much gold is there as opposed to how much might be) give it another twelve years of life—enough for it to notch up its 150th anniversary.

The unknown factor in Brazil is how much new gold may come from alluvial deposits in the vast Amazon basin that is now being opened up. There has long been considerable unofficial alluvial gold panning by *garimpeiros* on the rivers coming down from the Andes in Western Brazil and on various tributaries of the Amazon. The richest source of alluvial gold is currently the river Tocantinzinha, but extensive exploration for new alluvial deposits is proceeding on the Madeira River in Amazonas province. There is no precise estimate of how much gold the *garimpeiros* pan out every year, but it may well be 350,000 ounces. As a sign of the times the Bank of Brazil started buying gold officially from the *garimpeiros* in Manaus in 1972. The new discoveries are hardly likely to have any impact on the world gold scene, but they could lead to Brazil being one of the very few countries where gold production is actually rising. People there talk quite readily of the country's gold output doubling in a few years—something that one rarely encounters elsewhere.

IV AUSTRALIA TO ZAIRE

A casual glance at gold production figures for Australia in the early 1970s might suggest that another gold rush is on. Why else should output be doubling in a mere couple of years? Nothing so dramatic has happened since the original gold rushes to New South Wales and Victoria in 1852 and the stampede out to Kalgoorlie in Western Australia at the turn of the century. This time around, however, read

Rio-Tinto Zinc instead of old-time gold prospector. The source of this new fountain of gold is not some secret gold field in the outback, but the rain-soaked tropical island of Bougainville in the Territory of Papua and New Guinea, where Rio-Tinto has spent a formidable $400 million dollars developing a new copper mine. And the copper concentrate just happens to be heavily laced with gold. The Bougainville project, one of the most expensive mining ventures ever undertaken anywhere, is expected to yield a handsome dividend of between 500,000 and 600,000 ounces of gold a year; a bonus that alone should pay for the operating costs of the mine. This Bougainville gold will effectively step up Australian production from a modest 620,000 ounces in 1970 to well over one million by 1973. Australia is thus poised to leap up from seventh place in the world league table (behind Ghana and the Philippines) to fifth, right behind the United States.

The jump is all the more dramatic because it completely reverses the trend of Australian gold output which has been falling steadily since 1958. A decade ago Australia did rank fifth in the league, with production of over a million ounces annually. Since then rising costs and shortage of skilled labour have cramped the industry's style, particularly in Western Australia which has been the largest gold-producing state in this century (with a record 2·3 million ounces in 1903). The mines there, of course, draw little comfort from Bougainville, for their own situation has not changed much; Western Australia's production was down to a mere 350,000 ounces by 1971. Indeed, it is a dichotomy that at the moment Australian gold output leaps forward, the death of gold mining in its old form is still close at hand. The industry has been busy, in fact, writing its own obituaries for several years. As R. C. Buckett, president of the Western Australian Chamber of Mines, said gloomily at their annual meeting in 1970, 'Gold mining . . . is rapidly drawing to an end under the tremendous burden of rising costs.' Certainly the prospect then was a dismal one. The major gold mines of Western Australia—Gold Mines of Kalgoorlie, Lake View and Star, North Kalgurli (1912) and Central Norseman Gold Corporation—which between them provided half of Australia's gold were kept going solely by a subsidy of up to A$8 an ounce (U.S.$9·50). The subsidy, first started in 1954 under the Gold Mining Industry Assistance Act, buttressed the mines with nearly U.S.$4 million annually. Even that help, however, was beginning to wear thin. Lake View and Star and North Kalgurli said bluntly in 1970 that even with the subsidy they were ceasing development work and would mine only their known reserves. Gold Mines of Kalgoorlie and Central Norseman (subsidiaries of Western Mining Corporation) talked of ceasing operations within a year or two. A couple of the mines along Kalgoorlie's famous Golden Mile even began to switch their plant over from processing gold to nickel.

Faced with the imminent demise of its gold mines, the Australian government did step in with increased aid. From the beginning of 1972 the maximum subsidy was raised to A$12 (U.S.$14·28) an ounce; but, more important, the mines on subsidy were given a higher share of the benefits of free market sales of their gold. Initially all newly mined gold has to be delivered to the Reserve Bank of Australia and is paid for at the monetary price, with the subsidy added if the particular mine qualifies. The gold is then re-purchased from the Reserve Bank by the Gold Producers Association, a co-operative of all Australian gold producers, which is entitled to sell it on the free market. Until 1972 the Gold Producers Association had to pay the Reserve Bank three-quarters of the premium it gained on the free market, while the mines themselves got the remainder. This share-out was altered in 1972 so that the mines received half the difference between the official

Australian price and the free market price (in addition to the subsidy); in effect they are now getting the full free market rate. And the high price of around $65 in the latter half of 1972 did mean that some mines whose death warrants were being signed in 1970 were at least breaking even. Gold Mines of Kalgoorlie even restarted development work on new reserves. The real debate was whether they could be revitalised as long-term prospects. The trouble was that the running down of the industry in the 1960s had left it riddled with decay. The gold mines of Western Australia, in particular, were stuck with antiquated under-ground workings and outdated plant. Since everyone had been forecasting their imminent demise for years there had been no point in investing in new machinery. The gold prices of 1972 naturally reopened the question, but most mining executives felt that while a price in the $60 to $70 range could help the mines hang on for a while longer, it was not enough to warrant a massive modernisation programme. Furthermore these lonely mining communities have had great difficulty in wooing skilled labour from the cities, so that the high quality manpower that would be needed for expansion is hard to come by. The prospects for these famous old Australian mines, therefore, are still cloudy.

The future of Australian gold production really seems to rest on base-metal mining with gold as a useful by-product. The Bougainville project obviously is the pacemaker. The true size of this development can be appreciated when one realises that the yield of gold from this little island would alone retain Australia its present position in the world league table if all the other gold mines on the continent packed up tomorrow. Moreover, Bougainville was planned when the gold price was $35 an ounce; even at that level the gold was really just a fine dividend on a copper mine. Now every $1 increase in the gold price adds at least $½ million to Bougainville's annual income. It may be only a copper mine, but it certainly has a golden lining.

The pattern of gold as a by-product is already evident elsewhere in Australia. In Tasmania the gold comes from copper mining at Mount Lyell and lead-zinc-copper operations at Rosebery; in New South Wales it is a mixed with silver-lead-zinc concentrates at Broken Hill; and in Queensland the gold is in copper ore at Mount Morgan. This Mount Morgan mine is part of the Peko-Wallsend empire, which embraces no less than seven mines in Australia that produce at least some gold. Indeed, the Peko-Wallsend group ranked as Australia's number one gold producers in the fiscal year 1971–72 with 231,000 ounces. Their best mine is Juno, tucked away in the outback at Tennant Creek, about three hundred miles north of Alice Springs in the Northern Territory. Juno's gold is mixed with rich deposits of bismuth. In 1971 the mine, which is operated by a Peko-Wallsend subsidiary Orlando Mines, achieved the distinction of more than doubling its production to reach 180,000 ounces to become Australia's number one source of gold that year. Juno's reign as champion is likely to be short, however, for Bougainville will overtake it as soon as the latter gets into its stride.

The Philippines have been following this Australian pattern, although not so dramatically. But gold yield from copper concentrates has enabled the Philippines to jump up a couple of places to sixth position in the world league (above Australia until Bougainville comes in fully) in the last decade. Her production is now nearly 650,000 ounces a year, compared with around 400,000 in the early 1960s; nearly half this gold comes from the copper mines of Lepanto Consolidated, Philex Mining and Atlas Consolidated. The only true gold mining group of importance left is Benquet Consolidated, who have long been the major producers and still turn in around 250,000 ounces a year.

But Benquet and the other gold mines like Itogen-Suyoc, Paracale Gumaus and Atok-Big Wedge have been struggling, like gold producers everywhere, against rising costs set against the fixed price of gold. Since 1961 they have been given subsidies by the Philippine government, but this has barely kept them in the game. The maximum payment for large producers (Benquet) was raised to ensure a maximum of $48 an ounce in 1971, while smaller producers received up to $54. The even higher prices on the free market in 1972 were obviously a welcome shot in the arm, but hardly likely to alter the fact that the future of gold in the Philippines is really with the by-product producers.

The only other country in what one can call the major gold league is Ghana, which runs slightly ahead of the Philippines, with production of around 700,000 ounces a year. The major slice of this comes from the Ashanti Goldfields Corporation, which was taken over in 1969 by Lonrho, the London based investment firm. The Ghana government granted Lonrho a fifty-year lease in return for a 20 per cent stake in the company. In the rest of the world, however, gold production is still on an almost eighteenth-century scale, with a few thousand ounces here and a few hundred there. The U.S. Bureau of Mines annual survey reports gold production in over sixty countries on every continent except Antarctica. They even list 96 ounces for Cameroon, 36 for Nigeria and 35 for Mozambique in 1971, but all this only serves to emphasise the dominance of South Africa and the Soviet Union. Beside these giants, only Canada and the United States can muster one million ounces a year; only Ghana, the Philippines, Australia and Rhodesia top 500,000. Nine other countries—Brazil, Colombia, India, Japan, Mexico, Nicaragua, North Korea, Yugoslavia and Zaire—scrape home with over 100,000 ounces each (that is to say one three-hundredth of

South African output). The rest are really nowhere. In Western Europe, for example, France (as might be expected) is the champion, with a tiny 65,000 ounces a year; Sweden is runner-up with 45,000 ounces. There is no formal production in Britain, but some eager Boy Scout troops still make annual camping trips to the mountains of North Wales to scratch around for nuggets. If they are lucky they can make a few pence towards the costs of the expedition. The wedding rings of Queen Elizabeth II, Princess Margaret and of Queen Elizabeth the Queen Mother are all made from Welsh gold mined at Bont-ddu in Merioneth-shire in 1923.

The question for many people is how much does China produce? There is official silence on the Chinese gold front, but she is known to mine a little gold in the Liaoning and Heilungkiang regions near the Russian border. They are estimated to yield 50,000 ounces a year. But have there been any new discoveries? No one says anything. The Refinery of China in Peking is recognised as an acceptable melter and assayer of 'good delivery' bars, but its work has rarely been seen in the West. The only real sign of Chinese activity has been as buyer of gold in London in 1965 and 1967. Otherwise China remains an Oriental enigma.

6

London

Beneath the streets of the City of London on any weekday afternoon a handful of strong men in blue overalls can be discovered in the vaults of the five firms that comprise the London gold market busily weighing up neat piles of gold bars, packing them into light-weight boxes made of wood fibre and sealing them with metal strips. The men need to be strong. Humping around 400-ounce bars calls for muscle and stamina. While most banking transactions simply involve transferring figures from one column to another or sending a telex message, gold calls for hard physical work. As one dealer remarked, 'You've got to have good strong men and good strong scales.'

The scales of London's dealers have long weighed a major slice of the non-communist world's newly mined gold each year; indeed, until 1968 almost 80 per cent of all new gold passed through their vaults. Although the precise turnover of the market is a secret treasured as much as the gold itself, London has handled up to 1,000 tons annually for years, with records of 2,000 tons in such hectic years as 1967 and 1968. Most afternoons nowadays up to a couple of tons of gold vanish into wood fibre boxes and are whisked away by armoured van to London airport to all corners of the globe. Penetrating the security curtain of the vaults, one can see highly polished Russian gold, clearly emblazoned with the hammer and sickle, stacked side by side with oblong bars from the Rand Refinery in South Africa or squat, square American bars stamped by the U.S. Assay Office. Each bar must be up to the exacting 'good delivery' standards of the London market by being at least 995 parts per thousand pure gold, containing between 350 and 430 troy ounces of fine gold and bearing a serial number and stamp of one of the forty-nine firms in fourteen countries who are approved as melters and assayers of gold (they include the Refinery of China in Peking and the All Union Gold Factory in Moscow). Provided these strict requirements are met, the country of origin of the bars is not important, except that there is often a preference for Russian bars, since they are always 999·9 parts per thousand pure and this high quality attracts a premium price.

London, as the traditional wholesale market for gold, caters for all tastes and pockets. Amidst the heavy 'good delivery' bars in the bullion rooms one man may be busy packing small 10 tola bars (3·75 ounces) that look like luscious gold-coated chocolates and are the staple requirement of smugglers to India. Another may be checking 10 ounce bars destined for Malaysia or pocket-size kilo bars bound for Switzerland, France, Beirut or Kuwait.

London's role as the world's premier gold market for almost a century stemmed simply from the fact that right from the first discoveries on the Rand in the 1880s until 1968 virtually all South African gold was shipped there for marketing. Most Wednesdays a Union Castle liner left Durban with her strongroom stacked with gold bars; on arrival in Southampton the bullion was swiftly transferred to the vaults of the Bank of England, which acted as selling agent for the Reserve Bank of South Africa. Since South Africa produces over 75 per cent of the non-communist world's gold the London market, having cornered this golden torrent, was inevitably supreme.

Her pre-eminence naturally attracted other nations with gold to sell. The Russians, the Rhodesians, the Ghanaians and even the Philippines all sold much of their gold there too. This monopoly, however, came to an abrupt halt in March 1968 when the gold crisis shut the market for two weeks and led to the creation of the two-tier market, effectively segregating monetary and non-monetary gold. For a while South Africa sold no gold, because there were ample supplies from the vast overhang of 2,500 tons of gold bought by speculators during the crisis gold rush. Throughout 1968 and 1969 those Union Castle boats carried only a modest 400 tons in all, compared with their usual treasure chest of 1,000 tons a year. And when large scale gold sales did resume, most of the gold was air-freighted initially to Zürich. The South African government, somewhat disenchanted with Britain's Labour government's refusal to sell them arms, decided as a matter of policy that their gold should no longer go through London. They relented, however, in 1970 when the Conservatives won the general election; once again most of the gold came by sea from Durban to Southampton and into the vaults of the Bank of England. But there was one vital difference; the Bank was no longer acting as South Africa's agent in selling the gold—but merely providing its vaults as a kind of local warehouse. Moreover, the five members of the London gold market were only selling about 20 per cent (about 4 tons a week) of the gold. The rest was being sold by the three major Swiss commercial banks, the Swiss Bank Corporation, the Union Bank and the Swiss Credit Bank, even though it physically came to London. The reasons for this rather curious arrangement whereby gold for the Swiss banks came via London were primarily two-fold. First, it is cheaper to move gold by sea rather than by airfreight and the most practical way is to use the Union Castle boats to Britain, since they have strongrooms tailored for the job. But secondly, and perhaps more important in the South African decision, was the fact that the gold entering Britain, even if it is on account of Swiss banks, shows up in the

British import figures and is one way of clearly demonstrating the volume of South African sales to the commercial market. The Swiss do not publish figures on bullion imports, so any gold going directly there is not revealed. Thus both in 1970 and 1971 over 900 tons of South African gold came to London; of this the London market actually sold some 200 tons each year.

Although its long-standing monopoly of South African sales is no more, London still ranks as a wholesale gold market of immense stature. Its prestige all around the world as *the* leading gold market remains high. The price of gold in London is still the crucial measure for bullion dealers everywhere from São Paulo to Hong Kong and Beirut to Tokyo. Their business each day hangs on the arrival of the latest Reuters' message with the 'fixing' price.

While the discoveries of gold on the Rand in the late nineteenth century finally secured London's supremacy, the gold market had been significant for two hundred years before that and had grown in tandem with the City's overall expansion as a financial centre. Today there is a rare spirit of continuity stretching back to this early beginning in the firm of Mocatta and Goldsmid, which was founded in 1684, when Moses Mocatta set up as a bullion dealer in Camomile Street. Ten years later, when the Bank of England was granted its Royal Charter, Mocatta became its silver broker. Now, nearly three centuries later, Mocatta and Goldsmid, although a wholly owned subsidiary of Hambros Bank, is presided over by Edward 'Jock' Mocatta, a large, cheerful man, who is the ninth generation of the family in the business. The firm is still silver broker to the Bank of England, and the price of silver is 'fixed' every morning at 12.15 in their offices at Hambros Bank.[1] The bullion itself is

[1] In May 1973 control of Mocatta and Goldsmid was acquired from Hambros by the Standard & Chartered Banking Corporation in association with Dr Henry Jarecki, Chairman of Mocatta Metals Corporation in New York.

stored in the firm's old offices in Throgmorton Avenue
nearby, which have an atmosphere akin to the mid-
eighteenth century when Mocatta's operated from Grigsby's
coffee-house. The bullion is packed in a small room that, at
first glance, looks like a carpentry shop. The faded yellow
walls are hung with chisels, drills and saws. But the solid
wooden workbench is stacked with gold bars instead of
timber. An ancient hydraulic lift, operated simply by
pulling a rope to direct it, wheezes down through a hole in
the floor in one corner of the room into the depths of a
brick-lined vault below. There gold and silver is stacked in
neat little piles on wooden pallets. It is all so casual, that one
hardly seems to be standing in an El Dorado. Jock Mocatta,
tapping out his after-lunch cigar, says it is not quite so easy
for unwelcome visitors to enter. 'We have one or two little
secrets to stop robbers.' The family certainly should know
all the tricks of watching over gold after three hundred
years at it.

The second oldest member of the market is Sharps,
Pixley & Co., who, like Mocatta, are now part of a merchant
bank. They are in the Kleinwort, Benson, Lonsdale Ltd.
fold. Originally there were two firms, Sharps and Wilkins,
who began business as dealers in gold, silver and pearls in
1740 and Pixley and Abell, founded in 1852 by Stewart
Pixley, formerly a senior clerk in the cashier's office at the
Bank of England. The two firms merged in 1957 to become
Sharps, Pixley. Although the Sharps family connection
died out long ago, Stewart Pixley, the fourth generation of
the family to head the firm, took over from his father, also
Stewart in 1965. The senior Mr. Pixley, reminiscing shortly
after his retirement, liked to look back to the days before
the First World War when the gold market was a real
scramble. He remembered waiting for the South African
packet boat to arrive every Wednesday with gold from the
newly discovered mines. 'Brokers met that day,' he recalled,

'and fierce bargaining took place, although the price
never varied by more than a farthing.'

That aristocrat of merchant banking houses, N. M.
Rothschild & Sons Ltd., are the third oldest members of the
market. The Rothschilds have dealt in gold since the late
eighteenth century with the same combination of discretion
and flair that they have shown in such historic acts as
drumming up $11·2 million overnight for Disraeli to buy
Suez Canal shares. Nathan Mayer Rothschild established
the London offices in St. Swithin's Lane in 1804 and his
great-great-grandchildren hold court there today. The
Bank acts as host for and provides the chairman at the
twice-daily price-fixing ceremony.

Hard on Rothschild's heels came Johnson Matthey &
Co. Ltd., who officially date their foundation from 1817,
although the Johnson family were assayers of precious
metals as far back as 1750. They combine bullion dealing
with the establishment of a world-wide network of metal-
lurgical companies engaged in the refining and semi-
fabrication of precious metals. Six of the forty-nine companies around the world who are recognised by the London
market as acceptable melters and assayers belong to the
Johnson Matthey group, and they are the only members of
the London market with their own bullion refinery in
Britain.

The firm was originally based in Maiden Lane, but in
1822 moved to 79 Hatton Garden, which is still their
headquarters today. Since 1965, however, the actual bullion
dealing has been carried on by a new subsidiary merchant
bank, Johnson Matthey Bankers Ltd., in King Street, just
round the corner from the Bank of England.

Johnson Matthey have been deeply involved in every
side of the gold business for over a century. They conducted
many of the early assays for gold prospectors in South Africa
in the 1880s; several mining finance houses, including Gold

Fields, were floated on the strength of their reports. Before South Africa built the Rand Refinery in 1921 all her gold was refined in London by Johnson Matthey, Rothschild's (who had their own Royal Mint Refinery until 1968) and, until 1914, by a small refinery, Raphael's. More recently the group refined all Rhodesia's gold, until Ian Smith declared U.D.I. (Rhodesian gold now goes through the Rand Refinery near Johannesburg and is passed off as South African), and since 1970 have been handling most of the Philippines' gold.

While individual members of the gold market never reveal their turnover, Johnson Matthey certainly fabricate the major part of the 25 tons of gold used annually by jewellers and industry in Britain. In fact, their Hatton Garden office is one of the few places in the country where people can walk in off the street and buy gold over the counter. They must, of course, be authorised gold users because private citizens in Britain cannot legally hold gold. However, it is quite normal to see a young man in trendy gear wander in from the street, order a thin sheet of rolled gold and go off with it tucked under his arm.

Since Johnson Matthey are the sole member of the market to have their own refinery, the other members often assign them to make their small gold bars (the only alternative accepted melters and assayers in Britain now are Engelhard Industries Ltd., who are not members of the gold market). Indeed, the reputation of their bars around the world has been so high that they have frequently made bars for the Swiss banks, who find that a Johnson Matthey stamp may be a more readily acceptable passport to gold sales in places such as India than that of their own refineries.

The fifth member of the market are, by London standards, relative newcomers. As an official brochure puts it rather charmingly, 'The most recently formed member is Samuel Montagu & Co. Ltd. who began as a partnership in 1853.'

Although this may suggest that they are still serving an apprenticeship (everyone else, after all, has been going for more than 150 years) Montagu's, a leading merchant bank, are one of the most active dealers in gold. They have exceptionally good links with the Zürich, Paris, Toronto and Hong Kong gold markets and in 1963 pulled off the greatest gold coin coup for many years. They won the contract, in collaboration with the Bank of Nova Scotia, to replace the Uruguayan central bank's holdings of gold coin, worth between $60 and $80 million, with bar gold. Many of the coins turned out to be the rare U.S. $10 Eagle Liberty.

Their *Annual Bullion Review*, published each spring, offers one of the best analyses of the world of gold and silver for the previous year. Its columns also provide Montagu's with a useful platform to air their views on everything from smuggling to India to how much gold Russia really produces. They had a spirited argument in 1963 with the U.S. Central Intelligence Agency, when the latter first announced that estimates of Russian output should be drastically revised downwards. Since Montagu's had personally sold a great deal of Russian gold they felt entirely justified in sticking to their own estimates, built up from close dealings with Moscow.

Although a Montagu dealer may buy and sell a couple of million dollars worth of gold a day from his desk, with its battery of telephones and panel of push buttons giving direct lines to other members of the market, he rarely sees the gold. That is all handled deep in the vaults, where neat little piles of 'allocated' gold belonging to private clients are stowed away on shelves. Once upon a time the parquet flooring of the vault was ripped up and burnt once every twenty years to extract the precious grains of gold that had rubbed off bars into the wood. However, as this money saving exercise only produced $150 in gold last time,

Montagu's have decided that it is cheaper to forget about
the gold than buy a new floor.

The five gold firms are a very close knit fraternity. While
they can keep a few secrets from each other, they do like to
cloud themselves in mystery from the world at large. The
only visible evidence of their activity is the ritual of 'fixing'
the price of gold, which takes place at Rothschild's precisely
at 10.30 a.m. and 3 p.m. each working day.

The ceremony dates back to 12 September 1919.
Previously the Rothschild bullion broker had trekked
round the offices of the other gold dealers each day offering
South African gold for sale. But as the volume of gold from
South Africa increased, it seemed an easier, and certainly
more dignified way of doing business, to invite the other
dealers to Rothschild's each morning to fix the price for
the day. Until 1968 there was just a morning 'fixing', then
the afternoon session was added to coincide with the opening
of the New York market.

Moments before each fixing starts, four men walk up the
steps of Rothschild's marble offices in St. Swithin's Lane
and join Alan Jeffery, the manager of Rothschild's bullion
and foreign exchange department, in the gold-fixing room
on the first floor. The spacious room is fitted with an olive
green carpet, green chairs and an ancient pendulum clock.
On the walls are portraits of Francis I of Austria, Frederick
William III of Prussia and Empress Alexandra of Russia,
together with those of other European monarchs for whom
the Rothschilds negotiated loans in the early nineteenth
century. Alan Jeffery sits at one end of a long table with a
calculating machine in front of him. The other dealers sit at
four desks around the walls. On each desk is a telephone,
with an open line to the dealer's own trading room, and a
Union Jack. The flag enables a dealer to stop the meeting at
any point with a cry of 'Flag Up', while he confers with his
trading room, which in turn is in direct contact with a

dozen or so clients across Europe or in New York. Once he has their instructions, he lowers the flag and the ceremony proceeds.

The fixing actually begins with Jeffery suggesting an opening price, 'Gentlemen, we will start at $65·40'—or whatever may be appropriate that morning. The price will be based on the previous day's closing and any special new circumstances overnight. Jeffery and all the other dealers will already have talked to their clients before fixing and can gauge the mood of the day. (Up to 1968 all prices at fixing were quoted in sterling, even though the international gold business is always conducted in dollars. But with the introduction of the two-tier market, London sensibly switched to dollars.) Once the opening price has been made known, each dealer indicates if he is a seller, buyer or has no interest. A seller specifies how many 'good delivery' bars he is offering; a buyer merely states the fact, but does not reveal yet the number of bars he will take. If no seller appears at the opening price, then it is moved upwards initially by ten cents, and then in units of five cents until someone offers gold. Similarly, if there are no buyers at the opening figure, the price moves down. Once buyers and sellers have appeared Jeffery says, 'Figures, please,' and the buyers speak up. 'Mocatta sixty bars,' 'Sharps forty,' 'Montagu forty'. Then, provided there is enough gold for sale to meet the total demand, the price is fixed.

Over the years since 1919, a great deal of prestige and no little mystery has been attached to the five men with their little flags and telephones in the fixing room. Actually, the crucial decisions are not made there, but in the trading rooms of the five members, or by their customers. 'Our man at the fixing is purely a communications link,' said Jock Mocatta. Senior dealers rarely attend, unless there is some special matter of market policy to be debated afterwards, but it provides good experience for newcomers. One dealer generously offered to teach me in half an hour to represent his firm at the ceremony.

Until 1968 all business concluded at fixing was subject to a commission of a quarter per mille (25 cents on $1,000) —a rate which most dealers found absurdly low for it brought them only $250 on a sale of $1 million. Since 1 April 1968, however, the market has charged $\frac{1}{4}$ per cent (minimum $60) on all purchases at fixing; sales are effected free of commission.

There is, of course, no compulsion for a member of the market to do all, or indeed any, of his business at fixing. He is fully entitled to marry up all his own buying and selling orders where possible outside the ceremony. When Johnson Matthey, for instance, refined all Rhodesian gold they did not normally take that to fixing, they simply used it to meet their other clients' orders. And much of the actual dealing really begins once the fixing is over, with the price fluctuating during the day. Sometimes as much as 90 per cent of the market's business in a day is outside fixing.

But for many years the dealers had to rely heavily on fixing, because the Bank of England, acting as agent for the Reserve Bank of South Africa, was really their sole source of supply of new gold. The Bank would sell outside fixing, but was never very willing to do so. 'The Bank were the masters of fixing,' a dealer recalls. 'They were the only corner from which the gold came.' Although they were not personally represented there, Rothschild's acted as their agent and the Bank had an open telephone line direct to the fixing room. They held all the cards; besides the trump of being chief seller of gold, the Bank was playing several other hands as well. As the *Economist* once put it, they were darting 'into and out of the ring, donning four hats with the jerky animation of a hero in an early two-reeler.' The Bank was buying for Britain's own Exchange Equalisation Account, and, more important, was acting for the international gold pool consisting of itself, together with the central banks of Belgium, Italy, the Netherlands, Switzerland, West Germany, France and the Federal Reserve Bank of New York. A formidable portfolio, which enabled it to wield great influence in maintaining a steady price for gold. Today, it is quite extraordinary to look back on the mid-1960s, when the gold price moved only within the range of $35·00 to $35·20 year in, year out—the price at which the United States then stood ready to sell gold to central banks. The gold price throughout 1967, for example, moved only in the narrow range of $35·14½ to $35·19⅞. The moment it looked like going over $35·20, the Bank stepped in to sell from the gold pool to keep it down. And at fixing the bidding changed only by ½ a cent at a time; a price rise of two cents was big news.

Although those days are past history, it is worth looking at them in a little more depth to understand better where gold stands today.

When the London gold market first reopened after World War II, in March 1954, its activities were relatively peaceful for the first few years because of the shortage throughout the world of dollars, the staple medium of exchange for gold. Few European central banks or private buyers had any spare dollars to lavish on gold in those days. But once the flow of dollars out of the United States increased in the late 1950s, the world-wide demand for gold in a crisis picked up.

The first real test for the London market came in October 1960, when the forthcoming American Presidential election made many people wonder what the new President (John Kennedy, as it turned out) would do to solve the U.S. balance of payments deficit. Would he devalue? The uncertainty caused a flurry of gold buying. In London the price shot up late one afternoon to over $40, and looked like going higher overnight. The Bank of England held hasty conferences on the transatlantic phone with the gold and foreign exchange division of the Federal Reserve Bank of New York. They agreed that the Bank of England should make substantial sales to bring the price down.

The experience started the bankers debating how to draw up safeguards to prevent such drastic price rises in future. The Americans wanted to stop the drain on their own gold reserves, which had been built up to massive levels in the 1950s, but were beginning to decline sharply. The crucial thing was to hold the London price at under $35·20, because below that it was cheaper for central banks to buy their gold there, rather than applying directly to the U.S. Treasury for gold against dollars (the U.S. was selling at an even $35, but airfreight and insurance from the U.S. to Europe added 16¾ cents).

Accordingly, when the gold prices started creeping up again in the autumn of 1961, officials of the Federal Reserve Bank of New York flew to Europe to propose that they should join with the Bank of England and the central banks of Belgium, France, Italy, the Netherlands, Switzerland and West Germany in a sales consortium to sell on the London market in moments of crisis to stabilise the price. The Bank of England was assigned the role of operating agent. As the Federal Reserve's chief gold man said, 'It was like a Greyhound bus. We left the driving to them.' The Bank could draw on a pool of monetary gold provided in quotas by the member banks, with the Federal Reserve taking a 50 per cent stage by matching the combined contributions of the other nations. The scheme was given a pilot run in November 1961 and the gold price obligingly slipped down from the $35·20 mark where it had hovered all autumn.

The next spring, when the downward price trend continued, the Federal Reserve stepped forward with a second proposal: the gold pool should also undertake co-ordinated buying in London, whenever the price there was below the U.S. selling price (i.e. $35·16¾). Separate purchases of gold by the main central banks were thus replaced by the Bank of England buying for all and dividing the kitty every few months among the participants. The scheme was tested in February 1962 and confirmed in the April. Within a matter of months the Bank had recouped $80 million in gold for the pool. Then, in the autumn, the Cuban crisis sent the price soaring again, so the Bank weighed in with pool gold. They were unwittingly aided during that crisis by a sudden large sale of about 20 tons of Russian gold, which was made on the basis of instructions from Moscow that were clearly not co-ordinated with Khrushchev's trial of strength with John Kennedy. On the tensest afternoon of the confrontation the Russians suddenly offloaded enough gold to take the pressure off the price and the pool. Continuing large Russian sales that winter kept the price down and the pool soon lapped up $600 million in gold. This see-saw pattern continued for several years, although as the pressure of private buying for investment, industry and jewellery increased steadily from 1965 to 1967 (and Russian sales stopped), the pool began to lay out consistently more than it won back.

The devaluation of sterling in November 1967, however, really pitched the pool into its time of trial. The French were already playing awkward and opted out of pool operations shortly before sterling was devalued. The other members held on and after a meeting in Frankfurt declared that the combined resources of the pool (over $24,000 billion or about a third of all the gold in the world) could still maintain the gold price at $35. But the run on gold had started. On 14 December 1967 almost 100 tons of gold was bought in London (twenty times the usual amount). The pool then got both the London and Zürich markets to stop offering forward gold (a facility not resumed until 1972).

And President Johnson announced stringent measures to try to reduce the U.S. balance of payments deficit which was flooding Europe with more and more dollars. The market simmered down, but the pool had already laid out at least 1,000 tons worth over $1 billion in a matter of weeks.

The temporary calm ended abruptly early in 1968 when the Vietcong's Tet offensive shattered the illusions of anyone who thought that the Americans were winning the Vietnam War. Then General Westmoreland, the U.S. commander in Vietnam, asked for yet another 200,000 troops for the war. The danger signals for a continued heavy U.S. balance of payments deficit were flying again. The attack on the dollar was on. And the London gold-market was right in the centre of the battlefield.

Demand for gold rocketed. On Friday, March 8, it was again 100 tons, equivalent to all the gold mined in the Klondike gold rush. Over the weekend the governors of the pool central banks conferred in Basel. Late on Sunday night they stated: 'The central banks contributing to the London Gold Pool re-affirm their determination to continue to support the pool based on the fixed price of $35 per ounce.' And the chairman of the Federal Reserve Board, William McChesney Martin, vowed that the United States would defend the existing price 'to the last ingot'. In the ensuing week it looked very much as if the speculators meant to test him to that limit.

By Wednesday, 13 March, demand in London shot up to 175 tons, while Zürich handled over 80 tons and even Paris met orders for 15 (thirty times its normal daily turnover). Thursday the clamour was greater. London did at least 225 tons and the price had gone up to $35·25, despite the fact that the pool was laying out every last ounce of gold it could get hold of. An emergency airlift was even mounted by the U.S. Air Force to fly planeloads of gold from New York to the American airbase at Mildenhall in Suffolk.

From there it was hurried in heavily guarded convoys direct to the back door entrance of the Bank of England in Lothbury, which was closed off by police. (There is a nice story that an insurance agent travelling with the gold was so impressed with the security arrangements that he offered to reduce the premiums retro-actively.)

What the pool had not reckoned with, in supporting the London market with almost the entire world resources of monetary gold, was the extent to which countless speculators who had never even thought of buying gold before suddenly found it the fashionable gamble that week. Men who usually stuck sternly to cocoa or coffee futures got the word from their brokers that this was just not cocoa's week. Try gold instead. Many Americans, although they are not permitted to hold gold, bought through third parties in Zürich or Toronto. As *The Financial Times* columnist C. Gordon Tether remarked: 'It is common knowledge that it [the rush into gold] was in a major degree accounted for by activity of an unusually sophisticated character— activity involving funds of banking houses and other institutions that are normally not to be found taking any interest in the possibility of making a turn on gold.'

When the London market closed officially at 3.30 that Thursday afternoon, it was clear that very fast action was required to check the gold rush. The pool had already laid out nearly $3 billion in monetary gold (an eighth of their total reserves); both Italy and Belgium were anxious to opt out as their gold reserves dwindled. There was one immediate gambit to gain time—close the London market. A hasty Privy Council meeting was held at Buckingham Palace shortly after midnight and the Queen proclaimed Friday, March 15, a bank holiday. Britain's Chancellor of the Exchequer, Roy Jenkins, announced the news to a crowded House of Commons at 3.30 a.m. He noted that the decision to call the bank holiday and close the gold market had been taken 'at the request of the United States.' The Zürich market followed suit, but Paris stayed open and did hectic trading at prices up to $44 an ounce.

Meanwhile, with a little breathing space, the central bankers convened in Washington over the weekend and thrashed out an agreement which brought a new dimension to the world of gold. They announced an end to the gold pool and the introduction of a two-tier 'price' system for gold; the official price of $35 an ounce at which central banks would deal with each other and a free market price for the speculators, hoarders and industrial users, which would find its own level. The central bankers also said that they would not buy any more gold on the open market, thus, temporarily at least, sealing off their official reserves from newly mined gold. The two-tier system, in fact, divorced monetary from non-monetary gold.

The London market was ordered to remain closed for two weeks, while the precise details of the system were ironed out—an enforced holiday which several members resented, because it gave Zürich (which re-opened at once) a chance to set the pace in the new era. When London did begin again on April 1st, its whole character had altered. 'There has been a fundamental change,' a dealer of many years experience pointed out, 'The Bank of England are no longer the masters, they are a post office or a warehouse where gold is stored before it comes to the market. As for the price, that is highly volatile. We used to get excited if it changed 3 cents, now it can swing two dollars before lunch and no one notices.'

To begin with, the London price did not get markedly out of line with the official price of $35 an ounce. It hovered around the $40–$42 level during 1968 and 1969, and then at the end of that year, just as South Africa reached agreement with the International Monetary Fund about future sales (see Chapter 9), it tumbled right back to $35

and even under to $34·75. By this time, most of the specu-
lators of 1968 had moved out of gold, and the market had to
rely much more on the true demand of jewellery and
industry, which took the opportunity of the new low price
to build up depleted stocks. But by the end of 1970, with
much of that great speculative overhang of 2,500 tons sold
and the market relying increasingly on newly mined gold,
the price began to climb slowly. By the end of 1971 it was
almost $44; then, in 1972, it went through the roof. The
price cleared the $50 barrier at the beginning of May, but
did not pause for breath, and was comfortably over $60 an
ounce by June. For a heady day in August it touched $70
an ounce and then settled into the mid-$60 range until the
dollar devaluation of February 1973 sent it soaring beyond
$90. Montagu's *Annual Bullion Review* even forecast $100
price soon.

Such a market clearly bears little resemblance to the
sedate days of the 1960s, when it moved ½ cent at a time.
The dealers are clearly much more vulnerable with such
sudden swings, but it certainly adds more spice to the game
(and profit for the successful ones). However, London does
not have everything its own way now. The fact that 80 per
cent of South African sales go through Zürich has clearly
hurt their influence and their pride. Then the Russians, who
used to sell a great deal of gold through the Moscow
Narodny Bank in London up to 1965, switched their
outlet to Zürich when they resumed major sales in 1972.
Even gold from Ghana, the world's sixth largest producer,
now goes to Zürich instead of London, as part of a deal for a
Swiss loan. Furthermore, the five members of the London
market always act as individuals, unlike Zürich, where the
Swiss Bank Corporation, Union Bank and Swiss Credit
have had their own gold pool since 1968. The individual
firms in London cannot exert the same concerted influence
on the price that the Swiss pool can.

But if London is no longer omnipotent, it is by no means defeated. Its prestige is still high throughout the world; dealers everywhere wait for that vital telex of the level at which those five men with their little flags 'fixed' the price of gold each morning and afternoon. And many countries still buy their gold almost exclusively in London. A reputation built up over almost three centuries does not vanish overnight.

7

Zürich

Gold is as much a part of Switzerland as the Alps and
skiing. At the airports and the railway stations it gleams a
warm yellow greeting to the traveller from the counters of
the *bureaux de change*. Step out of a taxi or off a rumbling
tram in the Bahnhofstrasse in Zürich and there in the
windows of the Union Bank, the Swiss Bank Corporation or
the Bank Leu, gold bars and coins of all sizes rest seductively
on black velvet. At one bank a button in the lift even bears
the delightful inscription *'tresor'*. All this bullion so readily
for sale seems a sore temptation to the American or the
Englishman who is forbidden by law to hold gold. It is even
tax free. The amiable Swiss levy no tax on gold that is over
899 parts per thousand pure; all those bars in the banks are
at least 995. To the Swiss, of course, this is nothing to get
excited about. 'In Switzerland the traditional right of
every resident to acquire and own gold has always been
respected' said one bullion dealer. 'This has also called
for the metal being allowed to flow into and out of the
country without any restriction.' A colleague put it more
simply. 'Buying gold in Switzerland,' he pointed out, 'is
just like going to the baker to buy bread.'

Couple this ease of visiting the 'bakery' with the Swiss
banks' long-standing tradition of secrecy, numbered
accounts and anonymous safe deposits, and it is hardly
surprising that Switzerland is the largest market place
anywhere for gold. The banks in Geneva are busy with
clandestine shipments into and out of France, while those

in Chiasso on the Swiss–Italian border take care of Italians who prefer to buy discreetly just outside their own country. But it is Zürich that is the fountain-head of the gold. Zürich has been a major retail market in gold for a quarter of a century; since 1968 it has become the main wholesaler as well by the simple coup of luring most South African gold sales away from London.

The Swiss take gold and its disciplines very seriously. Until recently the commercial banks could even count their gold holdings as part of their liquid cash reserves. While the National Bank itself sets a good example by keeping at least half of Switzerland's own reserves in gold (they also have nothing to do with the newly created Special Drawing Rights). Their monetary gold stock, in fact, is the fourth largest in the world and is exceeded only by the United States, West Germany and France. No mean achievement for a country with a population of just over six million. Translated into the simplest terms this means that Swiss reserves contain about 13·2 ounces of gold for every man, woman and child in the land, compared to the United States which has just under 2 ounces per capita.

The Zürich gold market is dominated by the Swiss Bank Corporation, the Swiss Credit Bank and the Union Bank of Switzerland—the Big Three, in fact, of Swiss banking. Other commercial and private banks throughout the country naturally deal in gold bars and coins (the Bank Leu, for instance, is a major coin dealer), but almost inevitably they have to turn to these three leaders as their regular source of supply. As in London, the business is thus really controlled by a tiny group. 'Both markets,' Walter Frey of the Swiss Bank Corporation has pointed out, '[are] run by an exclusive circle of a few private enterprises to which outsiders have no access.'[1]

The Zürich club has been active in gold even since 1947.

[1] Walter Frey, *The Zürich Gold Market*, Euromoney, August 1971.

The London market was then still closed after the war and gold was in very short supply. But the Swiss Bank Corporation managed to scout out supplies from some central banks in Europe and even from the South Africans themselves. They found in those days that it was a better bet to change their dollars into gold, rather than Swiss francs, which were discounted 25 per cent against the dollar in the unstable post-war market. Initially the bank expected to keep the gold as part of its reserves until the Swiss franc strengthened, but they quickly found that the demand for gold in the Far East was so strong that it was more profitable to re-sell the gold swiftly. The greatest clamour came from China in the last days before Chairman Mao took over. Gold was selling in Peking and Shanghai for $50 to $55 an ounce and since it could be bought in Europe for $38 the traffic was highly lucrative. Looking back on those early days of getting a foothold in the Far East, one retired Zürich dealer told me, 'We made high profits and the middlemen between us and China made whopping profits. I had a visitor once who knew the Far East well and I sold him a substantial lot of gold. Years later he came in to see me again and he asked me "What did you make on that gold?" I said about 40 cents an ounce. He admitted he'd made $6 an ounce.'

Profits aside, those early deals in the 1940s enabled the Swiss Bank Corporation and the Union Bank and Swiss Credit, who quickly joined the game, to get to know local bullion dealers in markets all around the world. And those contacts have enabled them over the ensuing quarter of a century to build up formidable relationships, particularly east of Suez. Wherever one travels there, a bullion man from one of the Big Three in Zürich is sure to have been through the week before.

The reopening of the London market in 1954, however, faced Zürich with a considerable challenge. Quite apart

from London's own long-standing reputation and exclusive
connections with such key centres as Hong Kong, the South
Africans chose to funnel their gold to the reopened market.
Whoever gets that gold is king. Zürich's problem was that
it had no other major source of supply to fall back on,
except the Russians. But they too were selling chiefly in
London. So while London was the dominant wholesaler
during the next fourteen years, Zürich consolidated its
role as the number one retailer. A bullion dealer there
described their role rather charmingly as 'a turntable'
redirecting gold to buyers all over the world. As a retailer,
Zürich held several strong cards. The ease with which
anyone can buy gold in Switzerland and the freedom from
exchange controls naturally made it more attractive to
many private buyers than London. Investors from all over
the world were encouraged to keep at least 10–15 per cent
of their assets in gold. And many found almost magical
comfort in a nest-egg in a Swiss vault. The bankers are
understandably indignant at the frequent suggestions that
many of their clients are criminals or shady dictators who
use Switzerland as a hideaway for their loot. 'We are
bankers, not angels,' a manager at one of the Zürich banks
stressed to me, 'but we do try to do things right. We would
not knowingly sell gold to a gangster.' The key word, however, is *knowingly*; the gold dealers often do not know who
the true purchasers are, because lawyers or private banks
act on their behalf. But it would be quite wrong to imagine
that much of Zürich's gold is purchased to salt away the
profits of crime; such buyers would account for far less than
1 per cent of turnover.

The real point is that Swiss bank secrecy appeals to many
people, not just to criminals. American investors, for
instance, who may not buy gold privately, would be more
likely to make any under-the-counter purchases through
Switzerland, rather than London.

The bulk of Zürich's business, however, is not with clandestine gold hoarders. It is in the daily supply of gold to large consumers throughout Western Europe and around the world. Right on its own doorstep the Swiss watch and jewellery industry takes nearly thirty tons of gold a year; while simple geography makes Switzerland the handiest place for Italy, West Germany and France to purchase much of their gold. Since Italy and West Germany between them require almost half of all gold used in jewellery and industry in Western Europe each year, this in itself is a trump card up Zürich's sleeve. French hoarders, too, were major buyers of gold up to 1968, and since the legal import and export of gold in France was tightly controlled, it was smuggled in over the Alps.

But Zürich's arm goes far beyond Europe. They have fought a long battle with London for supremacy in the markets of the Middle and Far East, and in many of them have come out on top. In Beirut and Dubai, Vientiane and Singapore the Swiss have almost certainly captured much of the business from London. The great strength of the three Swiss banks in these markets is that they are operating as principals with their own substantial stocks of gold, whereas the members of the London market are essentially brokers. The Swiss, therefore, can afford to put much more gold into these markets on consignment and their prices are frequently more competitive than London's.

Although the Swiss Bank Corporation was first off the mark into the gold trade after the war, the Union Bank came in hard on their heels and really commanded number one position until 1958. Then the Swiss Bank Corporation made a comeback and have been jockeying with the Union Bank for the championship of the market ever since. The Swiss Credit Bank did not enter the fray in a big way until the mid-1960s, but have established good footholds in European and east of Suez sales. While all three banks are constantly active in overseas markets, Union Bank have also made something of a speciality of dealing with gold speculators and investors, whose gold, of course, never actually leaves their vaults. At the same time the Swiss Bank Corporation has been particularly busy in Latin American markets, where it has taken away considerable business from the Canadians, who had previously been able to supply those areas most cheaply.

These overseas forays have also been helped by the fact that all three banks now have their own gold refineries turning out an impressive range of small bars to cater to the latest whim of every market. The Swiss Bank Corporation owns the Métaux Précieux refinery at Neuchâtel, Swiss Credit Bank runs Valcambi at Balerna, while the Union Bank controls Argor in Chiasso. Their bars have undermined the almost exclusive hold which Johnson Matthey had on many markets a few years ago. In the early 1960s the Johnson Matthey small bars were the undoubted favourites throughout the world; every gold dealer from Beirut to Bombay and Manila to Madras knew and respected their stamp on a bar. A Johnson Matthey bar, in fact, was the American Express card of the gold trade. Bars of other refineries were often treated with some suspicion that they might not truly contain the right purity of gold. The small gold traders are very conservative fellows and a few had had experiences with gold bars with solid cement centres.

The Swiss, however, patiently campaigned for their own bars. The great gold rush of 1968 gave them unexpected help. In that flight into gold there simply were not enough Johnson Matthey bars to meet all requirements, so every possible refinery in Europe was pressed into service to turn out kilo bars for Europe and the Far East and ten tola bars for India. For a few months no one was fussy about whose bars they bought; the priority was to get gold in some shape or form.

Since then the Argor, Métaux Précieux and Valcambi bars have been readily accepted almost everywhere. Indeed the Union Bank had the satisfaction of finding that their Argor bars were preferred in Indonesia in 1969 and 1970 when that country was absorbing vast amounts of gold during the period of inflation following the fall of President Sukharno. The Argor bars just happened to be the ones first smuggled into Djakarta, so given a choice, many people continued to opt for them later on, because they were familiar with the brand image.

The real breakthrough for Zürich, however, was not so much in bars, as in persuading the South Africans to switch the flow of gold in their direction. Up to 1968, when all the gold went to London, Zürich was already buying up to three-quarters of all the gold offered there; and as the biggest customers in London they were inevitably able to dominate the price at fixing. Then the gold crisis in March 1968 gave them the opportunity they had longed for. When it was announced in Washington on Sunday evening, March 17 that the international gold pool was no more and that henceforward there would be a two-tier market for gold, the Swiss banks acted instantly. The managers of the bullion departments of the Swiss Bank Corporation, the Swiss Credit Bank and the Union Bank got together and decided not only that they would reopen for business next morning (while London was to remain closed for two weeks), but that they would combine to form a common gold pool.

They were thus able to set the pace for and capture much of the initial business in the new free market. Although they had no source of newly mined gold to draw on that was not important. Many of the speculators, who had snapped up more than two years' normal gold production in the preceding four months, swiftly liquidated their holdings, providing Zürich with ample supplies. The problem was not shortage of gold, but an over-abundance.

The new gold pool was designed to give more coherence to Zürich as a market and to help establish a standard Zürich price to match London fixing. The pool itself was not endowed with its own physical stocks of gold by the three participants. Instead, they committed themselves to go long or short for the pool's account, subject to certain (undisclosed) limits. Each bank, therefore, was still dealing independently in the market from its own stocks. During a day's trading each of the pool members is supposed to offset his individual market transactions by selling to the pool what he has bought, or buying from the pool to match his sales. The three dealing rooms are in constant contact over a conference telephone line, so that they can know from moment to moment the precise flow of gold into and out of the market and determine whether to raise or lower the formal pool price. This fluctuates by 5 cents at a time.

From its inception the strategy behind the pool was to present the Zürich market as a united front, rather than as three independent bullion dealers, in persuading the South Africans to sell their gold through Switzerland. The volume of South African production is so great that no single bullion dealer or bank can hope to handle it all. As Dr. Gerard Rissik, the former Governor of the South African Reserve Bank remarked to me in 1967, 'Our kind of gold knocks any small market for a loop.' The Swiss had to demonstrate to South Africa that they were not a small market; that they were just as capable of coping with 1,000 tons of gold a year as London. The pool helped give them that capacity.

Immediately after the 1968 gold rush South Africa did not sell any gold at all. She had a balance of payments surplus and could afford to bide her time to see how the two-tier market developed. The Reserve Bank in Pretoria, however, found Zürich (as well as London) bullion dealers on its doorstep eager to start selling gold when it was ready. At that time, as we have noticed in the preceding chapter, the South African government were disillusioned with the British Labour government's firm stand against arms for the Republic, so they were inclined to favour Zürich. Although the market there was still amply supplied with gold being sold off after the buying spree of March, a few test sales were made in 1968. The gold was airfreighted in great secrecy from Johannesburg to Zürich, and sold at a premium price.

Thus encouraged, the South Africans decided to entrust Zürich with much larger sales in 1969. The South African balance of payments was less favourable and they urgently required foreign exchange. The three Swiss banks, through the pool, committed themselves to sell a substantial proportion of South Africa's production (probably up to 1,000 tons) at prices well over $40 an ounce. It was a bold step and one which in the short term, at least, they came to regret. The bitter truth was that in 1969 the overhang in the market was still so large that it could not also absorb South African production. Gold buyers everywhere were biding their time while the price remained up over $40; jewellery manufacturers in Europe ran their stocks right down rather than buy more, and the markets of Beirut, Dubai and Vientiane were all quiet. The volume of sales was just too small. So Zürich got stuck with gold it could not sell. At the time the three banks put on a brave face and pretended that all was well. In fact, they all lost heavily. As one leading dealer conceded to me many months later, 'It's no secret. We lost a great deal of money. It was a very difficult time for us and South Africa.'

The Zürich dealers, more than anyone else, heaved a sigh of relief in December 1969 when the South Africans reached agreement with the United States that they could, in certain conditions, sell gold to the IMF. The agreement let Zürich off the hook from taking more gold for the time being. On the other hand Zürich's helping hand to the South Africans in 1969, when they really needed it, has had a long term dividend; they have been rewarded with the major part of South African sales since then in much more favourable market conditions. So although the pool nearly foundered in 1969 over its eagerness to tap that reservoir of South African gold, it has lived to see better days. At a press conference in July 1970, South Africa's Finance Minister Dr. Nicholas Diederichs, confirming that the bulk of the gold was sold through Zürich, added that one reason it still went that way was that the Swiss banks 'have helped us in the past.'

Although London came slightly back into favour in 1970, Zürich has continued to handle 80 per cent of South African sales. Thus in 1971 the three Swiss banks were responsible for the sale of at least 800 tons of South African gold, while London handled a modest 200 tons. The following year, with South Africa holding back about one third of her production, the volume through both markets was less. The extent of Zürich's continuing supremacy is often misunderstood because the bulk of the gold still goes by sea to England and therefore, apparently, to the London market. Actually most of it is destined for the Swiss banks, but South Africa prefers to send the gold openly through Britain, where it shows up in the import figures, rather than direct to Switzerland where it would not. Even then relatively little of the gold moves physically onward to Zürich, because it is cheapest for the Swiss dealers to direct it straight to their customers by air from Britain.

Its role as salesman for South Africa is not the only feather in Zürich's cap. When the Soviet Union began large gold sales again in 1972 much of it was channelled through the Swiss banks by the Russian bank Wozchod

Handelsbank, in Zürich. As an added bonus even Ghana switched her gold refining and sales from London to Switzerland, as security against a loan.

So Zürich is in the enviable position of having cornered well over half of all gold sales. In 1972 they handled the distribution of about 750 tons of the 1,200 tons of newly mined gold estimated to have come on the world market. That's not quite as profitable as inheriting a gold mine, but it may be the next best thing.

8

Beirut, Hong Kong–Macao and Singapore

I BEIRUT

Within the world of gold there has long been one city whose name springs most readily to the lips of customs men, gold dealers and gold smugglers at the first suggestion of contraband gold—Beirut. For over twenty years (from the 1940s to the late 1960s) Beirut enjoyed the reputation of being the prime launching pad for gold smugglers orbiting to the farthest corners of the globe, laden with 30 or 40 kilos of gold stuffed into the pockets of a canvas jacket beneath their shirts. Local dealers still love to recall the day no less than sixteen couriers in their golden corsets all trooped aboard a Pan Am jet bound for Hong Kong, carrying between them over half a ton of gold worth $500,000. There was some speculation on whether the plane would get off the ground; it did. The smuggling syndicates were such good customers of the airlines operating through Beirut, that there was considerable competition for their business; one calculation even suggested they spent $300,000 a month on airfares.

The evidence of their travels turns up in police files everywhere from Tokyo to Djakarta and New Delhi to Teheran. The Japanese authorities have dossiers on many Lebanese who were so enraptured with Japan that they came through as 'tourists' twenty times a year. They were not there, however, to see the cherry blossom or to dally with a geisha girl; merely to drop gold and leave again on the first available flight. And in Djakarta enquire how did gold smuggling (now operated out of Singapore) start there and someone says 'Ah, it was old so-and-so from Beirut.'

Nowadays Beirut cuts a slightly less dashing image. Many of the old smugglers have bowed out of the business and have retired with their profits to run hotels or night clubs; others have taken off for more lucrative pastures elsewhere Hong Kong, Vientiane, Singapore and Dubai have taken over many of Beirut's old territories; but the legends persist. Nevertheless Beirut remains the foremost entrepôt of gold for the Middle East, even if her lines of distribution are shorter. Her constant advantage is that the Lebanese government obligingly permits the free import and export of gold, while most of her neighbours, such as Turkey, Syria, Jordan, Iran and Egypt, do not. And since the Lebanese levy no tax on gold, the metal is relatively cheap in Beirut; normally only 20 or 30 cents above the London or Zürich price to take account of airfreight and insurance. So Beirut handles anything between 50 and 80 tons a year, most of it for re-export.

In the beginning of Beirut's heyday back in 1946, when gold prices fluctuated wildly, the business was highly profitable. Since the London market had not re-opened after the war, the Beirut dealers bought from Zürich. 'Those were the days when you could make as much as 30 per cent profit on a deal,' a banker recalled nostalgically 'there was a difference of $1 an ounce just between here and Kuwait.' Aging DC3s were chartered to fly 400-ounce bars from Beirut to the Persian Gulf where the gold was melted down into 10-tola bars for India.

One of the first firms to get into the market was Bullion Exchange Trading Company of Lausanne, which opened an office in Beirut in 1949 under the direction of a Frenchman, Paul Antoine Milhomme. For over twenty years, until Milhomme retired in 1969, Bullion Exchange controlled about a third of the Beirut sales. The Banque de Credit National S.A.L. and the Societe Bancaire du Liban were also important. They were joined by half a dozen local exchange dealers who have flitted in and out of the market.

Even after the London gold market reopened in 1954, Beirut remained essentially a preserve of the Swiss banks. Although individual members of the London market did make some sales to Lebanese banks and exchange dealers, and there was a strong preference at one time for Johnson Matthey kilo bars, the Swiss were always able to offer gold at more competitive prices. To begin with Union Bank of Switzerland was king of the market, but after some hard in-fighting in the early 1960s, the Swiss Bank Corporation gained a firm hold and won at least 30 per cent of the sales to Beirut. More recently the Swiss Credit Bank have jumped in to challenge them, operating through two newcomers in the local gold market, the Bank Saradar and the Advance and Commerce Bank. The task of the Swiss has become easier since 1968 as the long-standing preference for Johnson Matthey bars from London has declined, and the bars of Swiss refineries Argor, Valcombi and Métaux Précieux have become more acceptable.

Virtually all Beirut business is in kilo bars, because they fit so snugly into the pockets of smuggling jackets. And since 80 per cent of the gold that comes in will depart through unorthodox channels that is a major consideration. The gold that remains in Beirut is partly for the local jewellery business (which takes three tons a year), but primarily is used for another Lebanese speciality—fake coins. Originally the coin business was in Aleppo in Syria, just to the north, but the complications of shifting the gold bullion over the border and the finished coins back again eventually made it more practical to run the whole show from Beirut itself. So a couple of small refineries there churn out passable imitations of British sovereigns, American double eagles ($20) and eagle liberties ($10), French Napoleons, Turkish reshads, and even Mexican 50 pesos coins. A coin expert can usually sort the Beirut coins from the real thing, but to countless coins buyers around the world a sovereign is a sovereign and an eagle liberty is an eagle liberty regardless of where they were made. (Actually a good clue on the $10 eagle liberty is to look at the three arrows in a quiver on the lower right of the coin. The bottom arrow on a fake is usually blunter than on the real coin.)

Sovereigns have always been a hot selling line because the Moslem pilgrims to Mecca, who buy them by the score, insist on coins with a male head. Since the Royal Mint in Britain has made Queen Elizabeth sovereigns ever since 1952, Beirut obligingly fills the gap, stamping out George V and Edward VII sovereigns by the hundred thousand. Business is always especially brisk just before the *haj* pilgrimage season each spring as the coin dealers in Jeddah and other ports of Saudi Arabia stock up. Indeed, a small coin maker in Jeddah opens up just for the pilgrimage season, buying his gold either from Beirut or direct from London.

The five years of civil war in the Yemen up to 1967 also stimulated a strong demand for sovereigns, particularly among the Royalist Yemini forces. One Beirut dealer handled orders for over 6 million sovereigns for them during the struggle.

The demand for gold coins was also bolstered in the late 1950s by Arab sheikhs suddenly glutted with their first riches from oil. Some insisted on being paid in gold coin; a trunkful of sovereigns under the bed made them sleep soundly at night, with no bad dreams about the solvency of local banks. All the early payments to Saudi Arabia for her oil, for instance, were made in sovereigns from Beirut.

When King Saud was overthrown by his brother Faisal in November 1964, he was reported to have cashed in a treasure trove of two million coins on the Beirut market.

Today the oil sheikhs are much more sophisticated. They all have telex and play the markets of Europe with great expertise. They may dabble in gold from time to time if their Swiss bankers advise them to, but they are just as likely to get out of gold and into Euro-dollars if the return is more promising. When they do speculate in gold, they deal directly with Zürich rather than Beirut; the gold never leaves the Swiss strong rooms. Beirut is essentially a retail market these days for the jewellers and small-time hoarders of the eastern Mediterranean.

The precise flow of gold varies from week to week according to the prevailing mood of neighbouring governments and the customs authorities. If Syria suddenly tightens up, then the gold dealers in Beirut will immediately switch their activities to Egypt, Jordan or even further afield to Teheran or Kabul. If the Turkish border guards begin a special scrutiny of the trains coming through Syria from Lebanon, then plans can be made to send gold to Turkey by sea through the ports of Iskenderun or Mersin.

Year in year out Turkey is Beirut's best customer, absorbing at least half of all the gold going into Lebanon. The Turks are inveterate hoarders of gold; the city dwellers of Istanbul and Anakara buy Turkish gold coins, the farmers of the countryside put the profits of their harvest into 22-carat jewellery. Although there are strict controls on the import of gold into Turkey, the jewellers' shops which fill alleyway after alleyway of the Grand Bazaar in Istanbul are laden with golden necklaces, bracelets and chains. Beirut obligingly keeps them stocked with around 20 tons of gold a year. In 1970, just before the devaluation of the Turkish lira, Beirut could scarcely keep up with the clamour for gold; at least 40 tons found its way onto the Turkish market. And if one enquired in Istanbul how it got there, the standard reply was always, 'Oh, it comes in the stomachs of sheep and camels that stray over the border from Syria.'

Although large-scale movements of livestock are not evident between Lebanon and Egypt, plenty of gold also moves that way. The Egyptian peasant is as addicted to gold as the Turkish farmer; if he has a good harvest he puts the profit into gold jewellery and when his daughter gets married the event is crowned with gifts in gold. Since President Nasser forbad the free import of gold into Egypt as far back as 1956, the flourishing gold markets on the *souks* of Cairo and Alexandria have to buy on the sly. Beirut is naturally the most convenient source. It can be a lucrative trade. A kilo bar in Cairo fetches $300 more than in Beirut, which is a tidy profit of $9,000 on a standard shipment of 30 kilos; or perhaps $2 million on a year's throughput of seven or eight tons. Small wonder that some of those Beirut dealers have retired to open hotels.

Still they do lead a fairly demanding life. The gold dealer must be able to deliver to his smuggling clients at any hour of the day or night. Occasionally, when there has been a heavy demand for gold, dealers and smugglers have even rendezvoused at the airport to pick up gold from incoming planes from London or Zürich. As soon as the new stocks of gold have cleared customs, the boxes of golden kilo bars have been broken open in the boot of a car, and the yellow ingots speedily slotted into jackets. Then the courier is on his way. Although Beirut's smuggling heyday is over, the habit dies hard. As one dealer put it, 'Smuggling gets into the blood after a while. You can't sleep at night unless you've got a carrier making a run somewhere.'

II HONG KONG—MACAO

To the casual spectator lounging on the waterfront of

Macao's Porto Exterior, the white hydrofoil skimming in between the junks on the Pearl River is just completing a routine run in the hourly shuttle service between the tiny Portuguese colony and Hong Kong 40 miles away. But as the hydrofoil sighs back off its skis to halt alongside the Hong Kong Macao Hydrofoil Company Ltd. landing stage, a dozen policemen, who have been brooding sleepily in three dark blue Land Rovers for the past hour, tumble out and straighten their uniforms. All have revolvers at their hips, two also carry shotguns. The moment the hydrofoil is securely moored, there is a quick shuffling of papers between the captain and officials on the quayside. Then three Chinese coolies scamper aboard the hydrofoil and emerge a moment later, each staggering under the weight of four canvas bags tagged with the blue BOAC cargo labels. The bags are dumped on the floor of two taxis parked alongside the police Land Rovers and the coolies head back for more. Inside of ten minutes, forty bags have been loaded and the convoy moves away through the decaying streets of Macao with the taxis sandwiched between police vehicles. After a leisurely seven-minute drive they all pull up outside the peeling façade of the Cambista Seng Heng, a bank in the Avenido Almeida Ribiero and, under the watchful eyes of the police and the curious gaze of passers-by, the bags are quickly transferred from the taxis into the cool, dark interior. Then the metal grille gates clang shut. Another shipment of forty 'good delivery' bars of gold is safely lodged in Macao.

Among the gold markets of the world, the operations of the Hong Kong–Macao market are unique. They move along a narrow tightrope dangling between the legal and illegal, surrounded by a web of intrigue which seems to confuse even some of those involved in it.

'I defy God Almighty to come down here and find out who is in it,' says one local dealer.

The dual roles of these two colonies, hanging like ear pendants from the tip of China, arise because Hong Kong, although it has all the international sea and air connections plus the banking facilities essential for a gold market, as a British colony does not permit the free import of gold bullion. Macao is not hedged in by such restrictions. The market, which handles 50 tons of gold in a good year, therefore weaves together the respective advantages of each colony. Curious sidestepping takes place to keep within the law. The gold actually arrives in Hong Kong twice before going on to its eventual resting place in Japan, South Korea or Taiwan.

To begin with the gold is ordered through a consortium of two bullion traders—Mount Trading Co., Ltd. (owned jointly by Samuel Montagu the London bullion dealers, and Jardine Matheson, that pillar of Hong Kong's trading establishment), and Commercial Investment Company (associated with Mocatta and Goldsmid in London). The gold actually comes from Australia (Hong Kong taking most of Australian gold production), London and, occasionally, from Canada. Initially it arrives at Hong Kong's Kai Tak airport and is off-loaded into an armoured van under the close scrutiny of the Hong Kong Preventive Service. The van then speeds off to Kowloon docks, where the gold is stored in bond in a steel-reinforced concrete warehouse. There it waits until Macao's 'Sindicate de Ouro' is ready for another consignment.

The syndicate in Macao, which traded for many years under the more plebian name of the Wong On Hong Company, is the pivot of the whole market. Its founder and director was a wiry little man named Dr. Pedro Jose Lobo, who had a passion for gold and classical music—he conducted his own orchestra and even started a radio station in Macao to broadcast good music. Originally Lobo came from Portuguese Timor, where, as an abandoned child he had been adopted by a Portuguese family. He settled in Macao shortly before World War II, and over the years became the uncrowned king of the colony. He led an elegant existence in the 'Villa Verde' on a large estate surrounded by six houses occupied by his numerous children. In 1946, in his capacity as Economic Advisor to the Macao government, he started to organise smuggling to cater for a strong demand throughout the Far East for gold. His syndicate was granted a series of two-year contracts by the Portuguese authorities to have the exclusive licence to import gold. In return, the syndicate paid the Macao government an import duty on the gold, with a guaranteed minimum amount stipulated for each year. The gold tax soon became an essential part of Macao's income, accounting for at least a third of the colony's annual revenue. Exclusive contracts were also signed with dog-racing and gambling groups so that between them the three syndicates became the mainstay of the colony's economy.

At the outset Lobo had to fly his gold in via Saigon aboard an ancient Catalina flying boat, but in 1954, when the London gold market reopened, the Hong Kong authorities relaxed enough to allow the bullion to come through Hong Kong in transit instead. The Catalina kept up its golden airlift across the Pearl River until 1965, when the regular hydrofoil service between Hong Kong and Macao started. The syndicate did not, however, relinquish its control over the transport of gold, because they owned the new hydrofoil company. The profits of the gold trade enabled Dr. Lobo to found the Hang Seng Bank in Hong Kong, which came to play a major role in financing the gold traffic (the Hang Seng Bank is now partly owned by the Hong Kong and Shanghai Banking Corporation).

Dr. Lobo died in 1965, but his place at the head of the syndicate was inherited by a Chinese businessman, Y. C.

Liang, who was already one of the most eminent figures in the Hong Kong–Macao business communities. Besides heading the gold syndicate 'Y.C.' (as most people call him) was a director of the hydrofoil company, the Hang Seng Bank, the Macao Waterworks, and the Macao Electric Light Company; he also had interests in several large hotels.

The syndicate's operations in Macao, of course, are only half the story. Although the gold goes there quite openly by hydrofoil, it returns to Hong Kong in secret. Once those metal doors of the Cambista Seng Heng in Macao close on the gold from the hydrofoil it is never officially seen again. According to the export figures none of it ever leaves Macao—leading to the traditional joke: 'there are an awful lot of dentists in Macao.' Actually, there are a handful of Chinese refiners busy over their crucibles melting the big 400-ounce bars that come in from Australia or London into the 1, 5 and 10 tael (one tael is 1·33 troy ounces) ingots, which are the Chinese favourites for hoarding. These little bars come in all manner of convenient shapes to suit the most intimate needs of smugglers. The 1-tael bar can be bought either in a square like a solid gold postage stamp, or as a disc with a hole in the middle so that it can be slung round the neck or the waist on a piece of string. The tael bars are even made in a nugget shaped like a slipper bath, which can readily be taped to the navel or other more personal places.

All these small bars are swiftly smuggled back across the Pearl River into Hong Kong. A little gold does stay in Macao to be made into jewellery to tempt tourists strolling down the Avenida Almeida Ribiero on their way to lose all their money on the floating casino in the harbour nearby, but most is back in Hong Kong within three days of first going through there in transit. The contraband bars, in what one wag has christened 'a tael of two cities', are

spirited across the Pearl River by passengers on the hydro-
foils and ferryboats or in the scores of sampans and
motorised junks that ply the river bearing rice, chickens,
giant prawns and crabs to and from Canton. The Hong
Kong Preventive Service catch the occasional shipment—
just to demonstrate that they are on the job—but an un-
written and untalked about agreement between the two
colonies ensures that the flow of gold into Hong Kong is
not seriously interrupted.

A further paradox of the market is that once the gold is
safely back in Hong Kong, it becomes perfectly legal for
anyone to hold; only import is forbidden. Within the
colony, gold of 990 parts per thousand purity may be
exchanged freely. So no one asks any questions about the
origins of the tael bars in the vaults of the six main bullion
dealers who have their offices in the narrow, teeming
alleys around Mercer Street. In theory they have been in
Hong Kong since time immemorial; in practice they left
Macao yesterday.

The disadvantage of the Macao shuffle, however, was that
it did push up the price of gold. Until 1971 the Hong Kong
price was at least $3·50 an ounce over the London gold
price, of which 36 cents was airfreight and insurance from
London, $1·25 the Macao government's rake-off, about
50 cents for the syndicate's margin, another 24 cents for
smuggling back into Hong Kong and around $1·10 for
refining and other costs.

Even so Hong Kong offered the cheapest gold in the Far
East, and since the price was much higher in Japan and
other nearby countries, the Macao margin was not so
important. It was always enough, however, to tempt gold
smugglers to bring gold in direct from Beirut, Geneva or
Brussels by couriers on the commercial airlines. The
smugglers could undercut Macao and still make their own
profit. Since a good courier can carry at least 30 kilos of

gold, the gross profit on a single trip could be $2,700, given a $3 an ounce margin. Allowing for the airfares and courier's fee, that still left a handsome bonus.

Hong Kong rarely suffered, therefore, from any shortage of gold. Just for good measure the market latched onto a fresh caper in 1969, which brought in an additional flood of gold. Although the regulations forbid the import of gold bullion, they do allow the import of 14- and 18-carat gold alloys. A new Exchange Controller in the colony, anxious that the gold jewellery manufacturers there should be able to obtain their gold as cheaply as possible, agreed to permit imports of gold 'grain', skillets (thick sheets of alloy) and chain of not more than 18 carat. His approval was based on the calculation that it would not be profitable to remelt this commercial gold back into bullion. But he reckoned without Chinese ingenuity. The local bullion dealers found it was eminently worthwhile transforming the commercial gold into 945-tael bars. So quite suddenly everyone in Hong Kong was getting licences to import commercial gold; more than 40 tons went in both in 1970 and 1971. One enterprising merchant even got around the regulations to the extent that although he appeared to be bringing in gold grain of 14 carat, he was, in fact, importing a special concoction in which nearly 60 per cent of the pellets were *pure* gold and the remainder were *pure* lead; together, however, they were the correct weight for 14-carat grain. All he had to do was separate out the gold and lead pellets.

The Macao syndicate were most put out by this new traffic, for everyone in Hong Kong found it cheaper than buying from them. So after some discreet diplomatic string-pulling, the issuing of import licences was tightened up. The Hong Kong authorities do not like to rock the Macao boat, and, since it is partly the gold traffic that keeps Macao afloat economically, it is sacred.

Whatever way the gold comes into Hong Kong, the hub of the market through which it filters out all over Asia is the Chinese Gold and Silver Exchange Society at 14 Mercer Street. The exchange is open six days a week and its turnover on an average day is 20,000 taels, or 40,000 if business is brisk. Within the exchange there is a constant babble as the agents of the 192 members of the market shout into telephones positioned in niches around the walls (each telephone has a padlock to ensure it is used only by its rightful subscriber). Apart from the phones and a narrow wooden bench around the walls, the exchange is just a large open hall. A fan turns slowly on the ceiling, two or three members have their shoes cleaned as they wait for calls. The day's gold price is chalked on a blackboard in Hong Kong dollars, U.S. dollars, Philippines' pesos, Japanese yen and even South Korean won. The last prices are a sure guide to some of the destinations of Hong Kong's gold—for little actually stays in the colony, except to meet the demands of the local jewellers and hoarders.

Export, although not illegal, is discreet, because the smugglers obviously do not wish to advertise the fact that they are leaving Hong Kong with gold for fear the authorities at their destination might be tipped off. So the Chinese, who control the gold traffic in Hong Kong and everywhere in Asia east of India, have everything extremely tactfully arranged.

When Lobo first started his syndicate back in the 1940s the greatest demand for gold was right on Hong Kong's doorstep in China. During the years prior to the communist victory, the turmoil in China fermented the desire for gold; the junks heading up the Pearl River to Canton were laden with it. The best year ever was 1948, when the Macao syndicate moved over 100 tons of gold. Once the communists came to power, however, they effectively put a stop to gold buying by lopping off the head of anyone caught trafficking.

Since then there has been constant speculation as to whether or not gold is still being smuggled into China. The standard story is that it goes in to pay for opium smuggled out. But there is no real evidence of this, and most people close to the market in Hong Kong are confident that no gold is going in (although the Chinese government do buy from time to time in London). The argument in Hong Kong is that there remains a hoard of gold—along with diamonds and silver—hidden in basements and backyards all over China, but that no one dares to touch it. And it is highly unlikely that they would take the risk of buying any more; indeed, if anything, the gold is smuggled out of China to form a nest-egg abroad. There has been a steady flow of diamonds from China to Hong Kong ever since the communist take-over; some gold may have followed.

The closing of China's gates, however, did not alter Hong Kong's position as a superb dispatch point for contraband gold. The colony is astride the prime sea and air routes of the Far East; over 6,000 ocean going ships call there every year, while non-stop flights radiate throughout Asia. Moreover, Hong Kong has no exchange controls; one can walk into any money dealers with a bundle of Indian rupees, Vietnamese piastres or South Korean won and exchange them for dollars (albeit often at a terrible rate).

Hong Kong's best customer for years was India until Dubai, the little sheikhdom on the Persian Gulf, took over the exclusive supply lines in the mid-1960s. Then Japan's blossoming jewellery and industrial demand presented a profitable prospect. Officially the Japanese got their gold only from local mines, but these could not cope with requirements. The smugglers from Hong Kong obligingly made up the leeway. They were especially tempted because Japan maintained an artificially high gold price of $57 an ounce, when everyone else was still on $35. So even though gold in Hong Kong was around $39 the margin was a fat $18. Even better, the Japanese yen was a good strong currency, readily interchangeable into dollars. As Hong Kong's best customer, Japan probably took at least 20 to 30 tons of contraband gold annually between 1964 and 1968. One could often gauge on a particular afternoon that a secret shipment of gold was leaving for Japan that evening if there was a flurry of business in yen at the Gold and Silver Exchange.

Payment for the gold is sometimes more of a problem than smuggling the metal itself. Many smugglers from Hong Kong to South Korea, have found that their biggest headache—perhaps backache is the best word—is getting *won* notes out of Korea. They come only in very small denominations, so that a smuggler who is reimbursed in Seoul for 30 kilos, finds he has several suitcases full of won to take home to Hong Kong.

Despite its long-standing reputation as the Far East's premier gold market, things have not always gone Hong Kong's way in the last few years.

Since 1967 the authorities in Japan have been releasing much more gold officially, so that the amount required from smugglers has been less. And since 1 April 1973 the import of gold into Japan has been freed completely, so that anyone can buy gold there quite legally. Then a new gold market mushroomed at Vientiane in Laos to cater for the gold demand touched off by the war in Vietnam. Although most of the bullion people in Hong Kong had a cousin or an uncle in Saigon, they were unable to compete with the Vientiane gold price. The ultimate challenge, however, came from Singapore. The new gold market that started there in 1969 (see Chapter 8 section III) caught on so swiftly that it threatened to eclipse Hong Kong entirely. Singapore, quite simply, was offering gold at $2 to $2·50 an ounce cheaper than Hong Kong. The new market had the added advantage of being just next door to Indonesia which, in the grip of inflation in the late 1960s, was absorbing large amounts of gold. Up to 1969 Hong Kong had been catering to the Indonesian demand; Singapore took over.

Hong Kong thus entered the seventies with its gold trade in the doldrums. The commercial gold traffic had upset the Macao syndicate, and Singapore had snapped up one of her best customers. Clearly a fresh hand was called for if Hong Kong was to regain her former prominence. When the Macao syndicate's contract came up for renewal with the Portuguese authorities in 1971, Y. C. Laing decided it was time to bow out. The syndicate's profits were too small he felt to make it worth bidding for another two years. So after more than a quarter of a century the Wong On Hong Company gave up its Macao monopoly. And a new syndicate, christened Wo On Enterprises Ltd., stepped in. The newcomer, however, was no stranger to the gold business. Wo On Enterprises is actually owned by Cheng Yu-Tung, Czar of the Chow Tai Fook empire, one of the largest gold jewellery manufacturing and retailing groups in Hong Kong. Cheng Yu-Tung was full of optimism that he could revive Hong Kong's fortunes and combat the challenge of Singapore. Under his new contract the extra costs of diverting the gold through Macao were sharply reduced, with the authorities rake-off being cut to a modest 58 cents an ounce; other cost-cutting helped shave the difference between the Singapore and Hong Kong prices to a mere 30 cents an ounce early in 1972. For a few months Hong Kong thrived. Then the sudden sharp rise in gold prices in May and June 1972, stopped not only Hong Kong, but every other Far East market dead. For months not an ounce of gold went into Hong Kong or Macao. No one was willing to buy gold at $65 an ounce when it might drop to $55 the next day. So everyone waited. The Chinese can be a very patient people. 'They are worried by this sudden change,' a bullion dealer remarked, 'they will wait at least six months to see if it drops, and if it does go down, then they'll wait another six in case the slide continues.' Was it the end of the Hong Kong–Macao market? No, probably just a prolonged lull. There are still too many people throughout the Far East for whom gold is an essential hedge against evil times; a few 5-tael bars or even some 22-carat jewellery are a wise insurance policy. Everyone remembers the old Chinese maxim; '*Chin shih wan neng-te*— With gold anything is possible.'

III SINGAPORE

Most gold markets like to trace their history back proudly over many generations; in London Mocatta and Goldsmid have been going strong since 1684. So when some brash newcomer pops up and moves almost to the top of the world league in a matter of months the more traditional markets can naturally be a little put out. And no one has ever set people in the gold trade talking as much as Singapore did in 1970. It blossomed, like some fast-maturing lush tropical flower, from nothing into the Far East's biggest market in less than nine months, easily outstripping, and snatching the business away from Vientiane and Hong Kong. Local bullion dealers were to be found driving happily round in brand new Mercedes as they played host to bullion brokers from London and Zürich who came flocking out to see what the new market was all about.

The Singapore gold market, which officially came into being on April 1, 1969, was an essential part of the strategy of that forthright Prime Minister, Lee Kuan Yew, in turning his little island nation into a commercial and financial centre to rival Hong Kong. Singapore is not only within relatively easy reach by jet of London, Tokyo or Sydney, but is at the cross-roads of South East Asia; Bangkok, Saigon, Kuala Lumpur, and Djakarta are all within a couple of hours' flight. So the welcome went out to foreign banks and businessmen to set up in Singapore. They responded in quite staggering numbers; no less than 32 foreign banks from 13 nations had opened up in Singapore by 1972. The arrival of many of them coincided with the Singapore Monetary Authority (the equivalent of a central bank) launching both an Asian dollar market and a gold market in 1969. The conception of the gold market was unique. The official description called it a re-export market for non-residents only; that is a polite way of saying it was conceived as a handy jumping off point for gold smugglers bound for other Asian nations which prohibit the free import of gold. Gold could not be sold to the local people or even to anyone with a British or British Commonwealth passport; so citizens of Singapore, Malaysia or India could not buy, but visiting Indonesians, Thais, Japanese or Vietnamese could. Nor could the gold be paid for in Singapore dollars; it had to be paid for in a handful of specified currencies—U.S. dollars, Swiss francs, or German marks at the outset, Japanese yen were accepted later. Initially seven local banks and two bullion dealers were licensed to trade in gold, which came in—almost all in kilo bars—by airfreight from London, Zürich and sometimes Australia, on consignment from the three big Swiss banks and four members of the London gold market. The Union Bank of Switzerland found that the kilo bars of their Argor refinery sold particularly well. Since most people in Singapore were entirely new to the gold game, some of the licensed banks were slow to get into action. The Bank of Indo-China set the pace to begin with, but the Chinese-run United Overseas Bank and the British-run Chartered Bank eventually came out top of the market.

A complication to begin with was that the banks were required to deliver the gold kilo bars to their customers actually at the airport. This was an attempt to ensure that it really left the country and was not filtered instead to local jewellers (who get their gold under a separate quota system). But that regulation had to be quickly changed when it became apparent that most gold buyers were in fact leaving by sea (chiefly on the night boat to Djakarta) and had no wish to pick up gold at the airport. So the rules were relaxed to allow delivery to be made directly at a bank or bullion dealer's office, provided the customer (or his courier) showed his passport to prove he was truly a non-resident. Several banks then obligingly set aside a special room where their clients could stow the pocket-sized kilo bars in their briefcases, or, more often, in special canvas jackets beneath their shirts. The bank was then supposed to phone the customs to say they had sold the gold and occasionally a customs man would dash round to tail the gold courier to make sure he really left the country. However, by the time the bank telephoned the authorities their client was often already heading for the street, so that by the time the agent arrived he had vanished in the throng outside. The couriers, naturally, were not anxious to be trailed so sometimes four arrived at the bank at once, changed into different colour shirts inside (few people in Singapore's steamy climate wear coats) with the gold in smuggling jackets beneath, and then left the bank simultaneously by four different exits. If an unfortunate customs man was on lookout, he was either confused by the change of shirt or at best could trail only one courier.

Such tactics became even more complex in December 1971 after three couriers, each wearing a golden corset with 25 kilos of gold worth nearly $40,000, were murdered in a Singapore back alley and the gold stolen. They had been bound for Saigon, but were held up and knifed within minutes of leaving the bank.

Long before that gruesome hold-up, however, the Singapore market had got into its stride. Its real strength

was demonstrated at once; it offered the cheapest gold in
the Far East. The Singapore authorities simply took a
modest cut of U.S. $1 an ounce on gold going through,
while airfreight and insurance at Commonwealth rates
from London added only about 35 cents. So Singapore
could sell gold at about $1·40 over the London price; both
Hong Kong and Vientiane, the Far East's other distribution centres, were at least $3·50 over London. A kilo bar
of gold—the staple of the Far East traffic—cost $1,422 in
Hong Kong and $1,427 in Vientiane in September 1969,
while Singapore was offering it for $1,383. Naturally
everyone went to Singapore.

The new market held another trump card. Chronic
inflation in Indonesia during the late 1960s after the fall of
President Sukharno had created a substantial demand for
gold as the local rupiah became almost worthless. Before
the Singapore market was created on Indonesia's doorstep,
gold had been supplied by couriers working for European,
Beirut and Hong Kong smuggling syndicates. Singapore
was ideally placed for a take-over bid. Besides the opportunities offered by several flights a day to Djakarta, an
armada of ships plied out of Singapore to all the ports of
Java, Sumatra and all the other far-flung islands of the
3,000-mile-long Indonesian archipelago. Gold could leave
Singapore one afternoon and be in Djakarta, Surabaja or
Medan within forty-eight hours at the most. By happy
coincidence also the Chinese who run the local gold markets
in Indonesia (as everywhere else in South-East Asia) had
endless cousins or other relatives involved in the Singapore
gold trade; the Chinese connection made the pipeline easy
to establish.

By the autumn of 1969 Singapore was handling 4 or
5 tons a month. Most of it was ending up in Indonesia.
Some of it was quickly transformed into 100-gram bars by a
local refinery or into the lavish bracelets and necklaces that

adorned the scores of goldsmiths shops of Senen Market in Djakarta; but many of the kilo bars were simply salted away by hoarders anxious to protect their savings or hide the profits they were making from the new construction suddenly going forward as Indonesia picked up the pieces after Sukharno's fall. The gold trade was conveniently helped by the sudden flow of foreign capital into Indonesia to start these projects; more than \$1·6 billion was pumped in between 1967 and 1972 by the World Bank, American, European and Japanese developers. Some of it turned straight round to pay for gold. Despite the country's economic difficulties there was no exchange control. 'The phrase "exchange control" simply does not exist here' said a bank manager in Djakarta, 'if you have enough rupiahs you can come in to my counter and buy a million dollars.' Many did and traded the dollars for gold.

Singapore could not have opened its gold market at a more fortuitous moment. With the extra shot in the arm of a lower gold price by the beginning of 1970, when the London price was down to \$35 after being nearer \$40 for 18 months, it took off. Although there are no official figures on the market's turnover—the Singapore authorities do not wish to embarrass their neighbours in South-East Asia by revealing just how much gold their citizens are buying on the sly—at least 140 tons of gold went through in 1970. Indonesia took a good half; a nest-egg of some 70 tons, then worth almost \$80 million.

Singapore also wasted little time in establishing good connections with other local gold markets in South-East Asia. Bangkok, Saigon, Kuala Lumpur, and even Phnom Penh in Cambodia all found they could get gold from Singapore more cheaply than elsewhere. The chief sufferer was Vientiane in Laos, which had enjoyed a nice little corner for over five years as the chief supplier to its South-East Asia neighbours, especially South Vietnam. The trouble was that the dealers in Vientiane had to pay a 7·5 per cent duty to the government, which inevitably pushed up the price of their gold. Somewhat distraught, the Vientiane market dispatched a deputation to Singapore to ask if the Monetary Authority would not raise their $1 an ounce tax to enable the two markets to compete on a more equal footing. The Authority replied with a stern 'no'. Their new market was riding the crest of a wave; they saw no reason to jump off.

But this kind of surf-riding is always beset by many hazards. Singapore had no worries as long as Indonesia or South Vietnam required plenty of gold, and while the gold price remained in its trough of $35. By 1971, however, the wave was breaking fast. Quite apart from the rise in the gold price, the Indonesian economy was on a much sounder footing. Inflation there, which had been a terrifying 650 per cent in 1966, slowed to a mere 8·8 per cent in 1970 and an insignificant 2·4 per cent in 1971. So the rupiah became a more realistic currency. The commercial banks were also offering a tempting 24 per cent interest per year on savings account. Gold lost its magic. In 1971 Indonesia took only about 45 tons, chiefly in the first half of the year. By 1972 the traffic had ground to a halt entirely. Indeed, there was so much gold in Indonesia that several tons came back out to Singapore late in 1972, lured by the high international prices.

With its chief customer bowing out, the Singapore gold market looked somewhat embarrassed. They were not helped by the fact that up in Hong Kong a tough new syndicate had taken over the Macao contract (see preceding section) and vowed to get back what they had lost in Singapore's heady beginning. The fat on the Hong Kong price had been trimmed to bring it close to Singapore's. While up in Vientiane the Laotian's tax on gold imports had also been cut to make that market more competitive. Then the winding down of the war in South Vietnam also reduced that country's demand for gold. Singapore's problem was that it had no other strong clients to fall back on. So Singapore, sadly, seemed to be suffering the fate of all tropical blooms; they may blossom quickly, but they can also wilt fast in the heat.

9

Central Banks

The world's largest hoard of gold is not, as readers of *Goldfinger* might imagine, locked away in the bowels of the earth at Fort Knox. It rests 85 feet beneath the streets of Manhattan in the air conditioned vaults of the Federal Reserve Bank of New York at Liberty Street in the heart of New York's financial district. Anyone giving the bank forty-eight hours' notice can have a guided tour. The bank even has special booklets about the gold to hand out to visitors as souvenirs of their visit to this treasure trove, which was worth $15·5 billion in August 1972.

Although this is one-third of the non-communist world's official gold reserves, very little of it is American. It belongs instead to the seventy nations, who choose to keep part of their reserves 'earmarked' at the Federal Reserve in New York, rather than in their own central banks. Originally some of it was shipped over from Europe for safety on the eve of the Second World War. Most of it, however, has been purchased by European central banks since the late 1950s as persistent American balance of payment deficits brought U.S. reserves tumbling from a record $24·6 billion in 1949 to a shade under $10 billion by August 1971, when the U.S. suspended further conversion of dollars into gold. Since it is a costly business moving gold about, these central banks—with the notably exception of the French from time to time—leave it in the safe hands of the Federal Reserve, rather than go to the trouble of taking it home. The International Monetary Fund also keeps most of its gold at the Federal Reserve.

To view this El Dorado in Manhattan, the visitor walks in off the street and takes an elevator that slips swiftly down to basement level E, cut into the solid rock of Manhattan Island. There he passes through a series of corridors until he reaches a barred gate. A bank guard, with a .38 at his hip, reaches through the steel bars to inspect credentials, then unlocks this first barrier. Inside is a long narrow room in which bank officials, seated at desks, calmly shuffle papers commanding the movement of almost one-third of the monetary reserves of gold, soothed as they work by a non-stop serenade of piped-in music. Beyond them is the gold itself, shielded by a 90 ton revolving steel door, which is slightly cone shaped, so that at night when it is closed, it can be lowered three-eighths of an inch like a giant cork into a bottle. Bolts controlled by time locks slot into the steel, making the door completely airtight and watertight.

Inside this treasure chest, the gold is stacked neatly in 120 grey steel compartments, which are individually sealed and secured by three locks, whose keys are held by the bank's records department, the staff of the vaults and the bank's auditors. Each of the seventy nations whose gold the Federal Reserve guards has at least one of these compartments for its own exclusive use. The largest single compartment is stacked from floor to ceiling with 110,000 good delivery bars, worth over $1·8 billion. The bank, with tight-lipped propriety, declines to say whose gold it is.

When one nation buys or sells gold to another, the eight-man gold unit in the bank's foreign exchange department on the seventh floor is immediately notified by cable or telex. While the transfer is officially recorded, a note goes down to the vaults to switch the gold from one nation's locker to the other's. Normally four men working in two shifts are required to keep pace with the constant change of gold reserves. On a busy day they may hump up to 2,000 good delivery bars, each worth around $16,900 at the monetary price of $42·22 an ounce. So within a few hours of a transaction between, say, West Germany and the International Monetary Fund, the gold will have been physically moved from one locker to another. Since the lockers are not labelled with the name of a country, but simply with a number, a massive transfer can be made just by crediting the number to the new owners in the bank's books. Usually, however, transfers are small—often a bar or two to take account of interest payments—so good old-fashioned muscle is required, aided by fork lift trucks and conveyor belts, to hoist bars of gold to the top of a pile. The Federal Reserve proudly claim they physically handle more gold per year than any other institution.

In the last few years, however, life has been less hectic in these vaults and other depositories of monetary gold. Since 1968, when the central banks announced as part of the Washington Agreement that they would not buy any more newly mined gold for their reserves, the total monetary stock has been relatively static at just over 37,000 tons, worth $49·5 billion at the new monetary price of $44·22 an ounce. The only new gold to come into the system has been $780 million worth sold by South Africa to IMF in 1970 and 1971, under the terms of a special agreement reached in December 1969. Furthermore, as the gap between the monetary and the free market price widened in 1971 and 1972, most nations kept gold as a frozen asset in their reserves and made their international payments instead in dollars, sterling or the newly created Special Drawing Rights. And gold became even less mobile in the monetary system after 15 August 1971, when the United States refused to cash any more dollars for gold.

Although central banks sit protectively on their gold reserves, and gold remains an essential element of the international monetary system, its role has unquestionably diminished over the last generation. Before World War II to speak of a nation's reserves really meant its gold stocks; gold made up more than 90 per cent of the world's monetary reserves. Many nations, like the United States, kept all their reserves in gold. Today it is a vastly different picture; early in 1972 gold accounted for a mere 30 per cent of monetary reserves of $137 billion.

It is an incredible *volte face*. A few years ago gold was looked on solely as a monetary metal. Most newly mined gold went, more or less automatically, into the vaults of central banks, with a relatively small flow into jewellery, industry and private hoarders. From 1959, however, that pattern has been reversed; in every year since then, except 1963, more gold has gone into private hands than into monetary reserves. Over the decade from 1957 to 1966 about half of all newly mined gold went either into jewellery, industry or hoarding. In 1966, for the first time in history, *all* newly mined gold went into private hands and official reserves actually declined slightly as the international gold pool had to lay out gold from its stocks on the London market (see Chapter 6). That was but the beginning. After the devaluation of sterling in November 1967 the run on gold over the next four months, until the crisis of March 1968 forced the closing of the gold pool, drained nearly $3 billion out of official reserves. They dropped from $43·2 billion at the end of 1965 to $40·5 billion in mid-1968. And with only limited South African sales to the IMF since then, they had picked up less than $1 billion before the monetary price changed to $38 in 1971 revalued them upwards to $45 billion. In fact, official gold reserves finished up effectively all square in the decade 1961–71, while world trade, which reserves have to finance, more than tripled. The mainstream of gold has been diverted elsewhere.

For all that, gold is still the bedrock on which many central banks are founded. Some thirty countries (including Belgium, Switzerland and South Africa) still require by law that their central banks hold reserves of gold against note issues and occasionally even against deposit liabilities. Then there is the constant fear of devaluation of reserves held in dollars or sterling. France, Belgium and Holland learned this lesson in 1931, when sterling was devalued and they were caught with large sterling holdings. The experience has not been forgotten; the devaluation of sterling in 1967 and of the dollar in 1971 and 1973 were ample reminders. Of course reserves held in gold will not necessarily make money (gold earns no interest, unlike dollar reserves), but they are an insurance policy against losing it.

And there is a distinct psychological advantage for a country showing a healthy gold reserve. As Miroslav A. Kriz, vice-president of the First National City Bank of New York has observed, 'in countries whose populations have been, within a generation or two, victims of currency disorders, the showing of gold in a central bank's balance sheet is, at times, an essential monetary tactic.'

And R. C. J. Goode, president of South Africa's Chamber of Mines put it more bluntly at a conference on gold in London in 1972. 'I like having a hefty hunk of gold among my country's assets.'

Hardly surprisingly much of the pressure to demonetise gold has come from countries whose gold reserves are falling—and who are thus embarrassed by their shortage of gold. Would the United States be such a keen advocate of the phasing out of gold from the monetary system if she still cornered the major share of world gold stocks as she did from the late 1930s to the late 1950s? The phenomenon of the last fifteen years has been the shift in the balance of power, in terms of who holds gold, from the United States to Western Europe.

How did it come about? To begin with, from January 1934, when President Roosevelt raised the price of gold to the arbitrary level of $35 an ounce, until 15 August 1971 the United States Treasury stood ready to buy and sell gold to recognised central banks and monetary agencies at that price. This was the whole essence of the gold exchange standard; the dollar was readily convertible into gold. 'The dollar' as the saying went, 'is as good as gold.'

Initially, in the depressed 1930s, the United States benefited enormously. Not only was American mining stimulated by the price rise to achieve a record level of production, but other nations had little spare foreign exchange to buy gold. Fort Knox, that heavily guarded depository dug beneath the blue grass of Kentucky in 1935, began to bulge with American-mined gold, plus new supplies from Canada and South Africa. Leland Howard, former director of the Office of Domestic Gold and Silver Operations of the U.S. Treasury, once recalled that in the 1930s there was so much gold coming into the Mint to be remelted into good delivery bars, 'that we had it stacked out in the backyard like stovewood. In fact it was so plentiful we almost came to regard it as wood.' The river of gold continued to flood Fort Knox after World War II, when most central banks in Europe were forced to sell the last remnants of their gold for badly needed dollars to help rebuild their shattered economies. By 1949 the United States had amassed $24·6 billion in gold, almost three-quarters of the reserve of the non-communist world and virtually half of all gold mined to that date. Fort Knox was truly the greatest treasure trove on earth.

Fort Knox is not, however, the bustling bullion store many people imagine. There is no full-time staff humping gold bars around. The transactions of American gold, until convertibility was stopped in 1971, were handled by the U.S. Assay Office at Old Slip on the East River in New York, which kept its own working stock of $1 to $2 billion. It was from here, for instance, that gold was shifted by armoured truck the five blocks to the Federal Reserve Bank of New York if West Germany or France bought gold from the United States in the days of convertibility. Fort Knox was broached only when the Assay Office was either full of gold (in the 1930s and 40s) or running short (in the 1950s and 60s). Then, once or twice a year, the U.S. Mint, which is responsible for the Kentucky depository and the Assay Office, mounted a special gold-moving operation worthy of Ian Fleming's fictional raid in *Goldfinger*. A task force assembled at Fort Knox, broke seals, unlocked cages and either stashed away or hauled out at least 70,000 good delivery bars, worth $1 billion. Then the locks and seals were replaced and Fort Knox was left to its military guard for another year.

The problem for the Americans from the late 1950s onwards was that it was always a one-way traffic. The gold was going out of Fort Knox, never back in. And there was no secret of its destination; Western Europe. The grand reshuffle of gold reserves, with the United States as dealer, was on. Belgium, France, Italy, the Netherlands, and West Germany all began to replenish their gold reserves as their economies gathered strength. They were aided by the fact that the United States was running (and still is) a balance of payments deficit. The dollars flowed into Europe; the central banks there collected them up and came knocking at the door of the U.S. Treasury asking, as they were fully entitled to, for gold. The reversal of fortunes was clear even by the early 1960s. Consider, for instance, the transformation over a decade, of the official gold holdings of the following countries (in millions of $ at $35 per fine ounce).

	end 1953	end 1963
United States	22,091	15,596
Belgium	776	1,371
France	617	3,175
Italy	346	2,343
Netherlands	737	1,601
Switzerland	1,458	2,820
West Germany	325	3,843

Although it has been the French who have always made the most noise about the importance of gold as a monetary asset, the West Germans and the Italians quietly built up just as substantial holdings.

The rapid turn round by the early 1960s was just the beginning. Charles de Gaulle had yet to play his hand. The French kicked off 1965 with the cheerful announcement that they intended to convert a substantial proportion of their dollar reserves into gold. This pushed the price on the London market up from $35·11 to $35·20, which in those days was a major jump. The gold pool had to sell heavily to prevent the price soaring even higher. Then de Gaulle followed it up with his famous press conference at the Elysée Palace to an invited audience of a thousand journalists and diplomats. The gold exchange standard, he began, was no longer effective, a return to the true gold standard was essential. 'We consider,' he continued royally, 'it is necessary that international trade should rest, as before the two world wars, on an indisputable monetary basis bearing the mark of no particular country. What basis? Indeed, there can be no other criterion, no other standard than gold. Yes, gold, which never changes, which can be shaped into ingots, bars, coins, which has no nationality and which is eternally and universally accepted as the unalterable fiduciary value par excellence.'

In its long history no one ever gave gold a better testimonial. But de Gaulle's golden gauntlet was not received with enthusiasm in the United States and Britain, who were, after all, the prime targets of his wrath. James Callaghan, then Britain's Chancellor of the Exchequer, grumbled that the idea of returning to the gold standard was one of 'primitive barbarity'. But the French went on acting out the part. First they shipped back to France most of the gold that had previously been held to her account at the Bank of England and at the Federal Reserve Bank of New York. The sheer volume of the gold moved was staggering; from Britain alone over 1,300 tons—equivalent to an entire year's non-communist world production—was moved across the Channel in 1965. And a 100 tons of Russian gold was airlifted direct from Moscow to Paris, instead of coming to London, as part of a swop operation to avoid shifting even more.

Meanwhile the French were busy cashing in their official dollars for gold at the U.S. Treasury (as a member of the international gold pool, France could not buy for herself on the London market). They kept a mere 600 million in dollars to pay immediate foreign debts and to give the Bank of France some elbow room in the exchange markets; everything else went into gold. Between February 1965, when de Gaulle nailed his gold standard to the mast, and September 1966 France expanded her gold reserves from $3·7 billion to $5·3 billion. She held one-eighth of all monetary gold. Though the French made all the fuss, other European countries were quietly doing the same thing. In 1965 Western Europe as a whole switched $2 billion worth of dollars and sterling into gold. And by the end of the following year the scales had finally tipped so that the combined gold reserves of the six Common Market countries exceeded the dwindling American stock; the Common Market had just over $15 billion, the Americans $13·2 billion. If one adds in the reserves of other Western European nations, including Britain, the stockpile east of the Atlantic was over $22·5 billion.

To keep pace with this challenge, the United States was forced to take two important decisions to release more gold for foreign buyers. First, in 1965 President Johnson approved an act of Congress eliminating the traditional 25 per cent gold backing for Federal Reserve deposits. This freed $5 billion in gold to meet overseas demand. Then in 1968 Congress finally approved the removal of the 25 per cent gold backing for Federal Reserve notes; that sprung a further $10 billion and laid the entire remaining U.S. gold reserve of $13 billion on the line to defend the dollar.

But those persistent raids on Fort Knox could not be tolerated. The U.S. Treasury was already extremely reluctant to go on handing out gold for dollars and tried very hard to dissuade countries from doing so. Although in the last resort they had no option but to oblige with gold, they used every possible diplomatic and economic armlock to discourage the habit. As a Zürich banker remarked in 1966, 'The Federal Reserve had become rather like a guichet. You knocked on it, paid dollars and took gold out.' But by the late 1960s one had to knock very hard to make the clerk hear and he took extended tea-breaks to delay things. Then, on 15 August 1971, he finally went on strike. The golden guichet stayed closed; America would no longer trade dollars for gold. The suspending of convertibility, as part of President Nixon's package to restore health to the American balance of payments, was the end of an era. The gold exchange standard was no more. Although the Americans have talked about restoring convertibility 'in the future', that day hardly seems close.

By the time the guichet shut America's gold stocks were hovering around the $10 billion level, which had always been regarded as a psychological barrier. Pressures from the French and Swiss in 1971 to change more dollars into gold threatened to break the barrier.

The actual state of play in 1971 showed how commanding the Western European position had become; the gold stock there was over $20 billion, against $10 billion in the United States. Other developed countries such as Canada, Japan and South Africa accounted for a further $2 billion and the International Monetary Fund for $4·7 billion. The rest of the world (excluding China, the Soviet Union and her associates) mustered only $3·6 billion.

Throughout the years of raids on U.S. gold stocks only two major nations, Britain and Japan, had really played the game according to the American rules. The British, with sterling under constant attack in the mid-1960s, were in no position to buy gold. Britain is the lone European nation whose gold reserves have declined substantially over the last twenty years. In 1953 Britain had $2·3 billion in gold (over 70 per cent of her reserves), which represented the second largest stockpile anywhere outside the communist camp; by 1971 her gold holdings were down to under $1 billion, or less than 10 per cent of her reserves. The Japanese, as part of their post-war rapprochement with the United States, always obliged the Americans by rarely trading in dollars for gold. They consistently kept the major part of their reserves in dollars and held only a token stock of $500–$600 million in gold.

The one institution whose gold stocks have soared substantially in the 1970s has been the International Monetary Fund. Previously their stocks had fluctuated for almost twenty years at around the $2 billion mark, but they doubled from 1970 to 1971 because of increased quotas from members and, more important, South African sales. The IMF is now the only channel through which South African gold can be purchased openly for monetary purposes. The IMF has assumed, therefore, additional significance in the world of gold. After the revaluation of the metal at $38 an ounce in 1972, their stocks stood at $6 billion, or 14 per cent of all monetary gold.

The IMF was originally born out of the 1944 Bretton Woods Conference, which was convened to discuss the future of international monetary co-operation when peace was achieved. The Fund actually started business at its headquarters in Washington D.C. in the spring of 1947. By 1972 it had 124 members. Each new member, to establish his credit with the Fund, makes an initial deposit or quota and can then borrow double that amount. The size of a deposit clearly depends on the relative wealth of each newcomer, but under normal circumstances at least 25 per cent of the quota must be in gold (this is called its 'gold tranche'), the rest is usually in its own currency. The gold is held for the Fund by the Federal Reserve Bank of New York, the Bank of England, the Bank of France and the Reserve Bank of India in New Delhi.

Gathering a gold tranche nowadays, however, can be a tough business. With the high free market price for gold, central banks have become allergic to trading off any of their precious monetary gold to newcomers to the Fund to enable them to put together their gold portfolio. In 1971 Syria had to go against the rules of the Washington Agreement and buy gold on the free market to meet its gold quota at the Fund because no central bank would supply it. And in July 1972 France and Japan refused to take part in an international operation mounted by the IMF to raise a modest $7 million in monetary gold to pay the gold subscriptions of three new members, Qatar, Bahrein and the Union of Arab Emirates. Bangladesh had similar problems, until the Canadians generously stepped in to sell them $2 million worth of gold for sterling. Even then the Fund had to allow the new state an exceptionally low gold quota.

Gold deposited with the IMF does not necessarily lie idle. The Fund sells it to acquire currencies for its members. West Germany, for example, replenished its gold stockpile partly by buying gold from the Fund in return for contributing marks to support operations for sterling in the mid-1960s.

But the IMF achieved new significance after the Washington Agreement of March 1968. To the South Africans, shut off by that agreement from selling any more gold direct to central banks, the Fund represented the only alternative customer if they could not sell all their gold on the free market. The IMF Articles of Agreement stated that any member could sell gold to the Fund, including gold from local mines. But until 1968 the IMF had never actually bought much—its gold had simply come from members' subscriptions. Initially the United States tried to block South African attempts to sell to the Fund, much to the annoyance of the IMF staff and its managing director Pierre Paul Schweitzer. But President Nixon's new Secretary of the Treasury, David Kennedy, taking office early in 1969, decided that the U.S. relations with the Fund and South Africa had to be smoothed out. In December that year the U.S. Under-Secretary of State at the Treasury Paul Volcker met South Africa's Finance Minister Nicolas Diederichs secretly in Rome to hammer out an understanding to allow South Africa to sell to the IMF. They agreed that South Africa could sell a restricted amount of gold to the IMF when the price on the free market was at or below $35. And regardless of market price, South Africa could sell to the Fund when experiencing a balance of payments deficit. The Republic was thus in a position to take advantage of premium prices in the free market, but also benefited from the establishment of a floor of $35 to the price (later raised to $42·22 by the dollar devaluations of 1971 and 1973).

The Rome agreement came at a crucial moment for South Africa, which was heavily overloaded with gold and suffering a balance of payments deficit. She was able to dispose of $640 million to the Fund in 1970 and a further $140 million in 1971. Since demand on the free market picked up also in those two years she succeeded in selling all new production and worked off some of the backlog from 1968 and 1969.

While becoming a new depository for South African gold, the Fund was also busy presiding over the creation of the metal's new rival, the Special Drawing Right (SDR), the first attempt to create 'paper' gold.

The SDRs grew out of a fistful of plans in the 1960s to develop a new reserve unit, acceptable to all members of the IMF, that would increase world liquidity which was then more or less static. The concept was not new. Lord Keynes had proposed his 'bancor' back in the 1940s. His proposition had been that international reserves should consist only of gold and bancor; gold being used to buy bancor, but bancor not being reconvertible into gold. In those days, however, the Americans held all their reserves in gold and were sitting on their huge stock of the metal. So they showed no interest. Circumstances in the 1960s, however, forced them to look at the chances of making 'paper' gold.

The real breakthrough came in August 1967 when the finance ministers and central bank governors of the Group of Ten (the major developed nations of the West) thrashed out a common policy in London. After a final twelve-hour marathon session they announced that a new reserve asset to supplement gold, dollars, sterling and existing IMF credits would be launched. The new reserve asset would be administered by the IMF and strictly related to existing IMF quotas. It was christened Special Drawing Rights. What is an SDR? In the first place it is a credit unit, it is not a standard of value. Gold remains the standard of value and SDRs are defined in terms of gold. An SDR unit is stipulated to be equivalent to 0·888671 grams of fine gold. Each member of the IMF has SDRs allocated to it according to the size of its existing quotas and once doled out they become a permanent part of the world's reserves. Although they are related to gold they cannot be traded in again for gold at the IMF. Once issued they can be used by each country according to need, either being kept as monetary reserves or used to meet overseas payments.

The new unit was formerly born at the beginning of 1970, when SDR 3·5 billion were distributed, as the first stage in a three-year plan to hand out SDR 9·5 billion. The United States, as the biggest quota holders with the IMF, was the chief beneficiary each year. At the final allocation of the initial package in January 1972, for instance, the U.S. received SDR 710·2 million, Britain SDR 296·6 million, West Germany 169·6 million, France SDR 159 million, Japan 127·2 million and Canada SDR 116·6 million. Other members of the IMF received smaller parcels.

So far the SDRs represent a very small part of world reserves (after three years only one-fifth as much as gold), but the important thing is that they have been created at all. No one expects or intends them to replace gold in the immediate future. Despite considerable scepticism about their usefulness, many of the Fund's members employ the new units quite actively. They have traded the SDRs among themselves, and back to the Fund itself for currencies. Transfers in the first two years amounted to $2·3 billion. The Fund has even found that several members are prepared to be re-imbursed with additional SDRs instead of gold, when it buys currencies from them.

The Fund has also encouraged its members to 'think in SDRs' instead of dollars. To set a good example the Fund began to shift its own statistics over to an SDR standard in 1972. In the past all IMF figures had been in dollars, but after the devaluation of the dollar, the Fund made haste to express all its lending and repurchase transactions in SDRs. Thus American reserves, for instance, were quoted early in 1972 as SDR12·8 billion, instead of $13·9 billion.

For all this encouraging start the SDR is in its infancy, and it does not, of course, embrace countries such as China and the Soviet Union which are not members of the IMF. For them gold or currencies are still the key to international payments. Russia has long relied on gold sales in its trade with Western countries. Although it is now much easier for the Soviet Union to get extended credit terms, the major purchases of Canadian and American wheat announced in 1972 were inevitably coupled with heavy gold sales to earn the required foreign exchange. That gold, of course, did not accrue to central banks in the West; it was all sold on the free market. The Russians, however, underlined their faith in gold for monetary purposes in a rare pronouncement by a senior civil servant in their Ministry of Foreign Trade. In an interview in 1972 with the Cuban News Agency, Prensa Latina, Sergei Shevchenko was reported to have said that the Soviet Union did not feel that one currency or even several currencies could be the basis for international trade. 'World trade,' he stated, 'must rest on a more solid foundation, and that solid foundation can only be gold . . . gold should be the basis of everything.' A worthy successor to Charles de Gaulle.

While Russia, as a major producer of gold has been a heavy seller on world markets, China has stepped in from time to time as a considerable purchaser. In 1965 (at the same moment that France was making her big gold purchases), China bought 100 tons of gold worth $110 million on the London market; the following year she came back for another 30 tons. Two years later China topped up with another 60 tons. The main reason behind these forays into the gold market appears to have been to divest herself of sterling. Although no official figures of China's reserves are available, it is likely that she substituted a good part of her holdings of sterling for gold before devaluation in 1967.

Since 1968, however, the Chinese have been lying low and do not appear to have either bought or sold gold. Perhaps they are sticking to the spirit of the Washington Agreement that no new gold should be taken into the monetary system, even though they were not parties to it.

How faithful other central banks have been to that agreement is uncertain. Portugal, the Congo and some other small nations are known to have flouted it by buying direct from South Africa or the free market for their reserves. While the high price of gold on the free market in 1972 tempted the central bank of Uruguay to sell 33 tons on the free market to settle overseas debts.

Yet for all these difficulties, the divorce—trial separation is, perhaps, a better description—between monetary and non-monetary gold since the creation of the two-tier market in 1968 has worked reasonably well (certainly much better than many people predicted). The system has even stood the test of a $50 and more difference between the monetary and non-monetary price. But if the two-tier system has led to the freezing of monetary gold in the reserves of the central banks, it has created, on the other hand, a bustling scene in the private sector.

10

Private Buyers

On the morning of Thursday, March 14, 1968, an eminent European business executive, who normally resides in Switzerland, was having breakfast in the dining room of a hotel in the West End of London. As he ate, he looked over the headlines in the London morning papers: 'The rush for gold goes on,' said *The Financial Times*; 'London gold market hits all-time peak.' reported *The Times*. They made highly disturbing reading to a businessman heavily involved in international trade. He made some hurried phone calls to one or two economic experts he knew in London, then jumped into a taxi and headed for the offices of a member of the London gold market, just round the corner from the Bank of England. There he quickly bought $250,000 worth of gold. It was something he had never done before—he normally specialised in the commodity markets for copper, tin and rubber—and might never do again.

As he was buying in London, many other speculators—some with a couple of thousand dollars, some with a million—everywhere from Frankfurt, West Germany to Houston, Texas were joining the undignified and somewhat panic-striken scramble for gold. As early as 7.45 that morning a Swiss bullion dealer was yelling exasperatedly into the phone, 'No sir, I can't guarantee that we've put your order through to London.' In Paris, the police were out in force around the Bourse, trying to bring some semblance of order to a near-hysterical crowd of potential gold buyers who were clamouring for kilo bars at prices 20 per cent above the normal level. An old *commis de change* said he had never seen anything like it in all his eighty years.

Hardly surprising, for in what may well have been the last great gold rush of the common man, more gold was being traded every day than was mined in most gold rushes of the nineteenth century. By the time the markets closed that Thursday evening the total of private purchases in less than a week had been a good 1,500 tons; adding in the buying over the preceding four months, some 2,500 tons of gold had passed into private hands.

This stampede into gold, touched off by the belief that it was about to be revalued, broke all records. But gold has traditionally been the lifebelt that people clutch in a moment of panic. Fear of devaluations, political upheaval, or war has sent them scurrying into gold for generations. Gold is the talisman, the friend for all seasons. Countless men and women owe their lives to having a small bar of gold or a coin at the right moment. Many Austrians and West Germans escaped being sent east by the Russians in 1945 by slipping the soldiers gold coins. In Greece and in France there are entire families whose life savings were wiped out because they did not take the precaution of turning all their savings into gold in 1939 and hiding it in the cellar or in the vegetable patch. 'Everyone in Europe liked to have a little hoard, in case he had to run away,' recalled a London dealer.

The son of a wealthy soap manufacturer from Salonika in Greece once told me how his family changed all their money into sovereigns in the winter of 1941, just before the German invasion. 'We had at least 3,000 sovereigns hidden behind the frames of four doors in the house,' he remembered. 'Every time the doors slammed in the wind my mother used to come rushing out of the kitchen in case the sovereigns had been dislodged. As soon as the Germans arrived they took over my father's factory and without those sovereigns we would all have starved. Once a year we took out a door frame and removed a few sovereigns to keep us going. Most of our friends and relatives did the same thing. But my grandfather, who put his faith in Greek currency, was left with a bundle of worthless notes. He lost everything.'

This experience dies hard. Although in Europe, the security of the last generation is gradually weaning people away from the habit of keeping a little gold under the bed, it does not take much to provoke them to return. The fear of war may have vanished, but the threat of devaluation and inflation continues. And if the traditional peasant hoarder now prefers to invest in a television set rather than a kilo bar, there are plenty of multi-national companies who like to keep a little reserve in gold, against a rainy day.

Outside Europe, the hazards of political and economic crisis regularly trigger off gold buying sprees. When Colonel Gaddafi expelled many Italians from Libya after his coup in 1969, the only way most of them could get money out was in the form of 22-carat gold jewellery; the *souks* in Tripoli and Benghazi enjoyed record sales. Asians living in East Africa have always put most of their savings into gold (the African himself prefers to buy an extra cow), knowing that one day they may be thrown out; a wise precaution as the actions of General Amin in Uganda in 1972 demonstrated. In Turkey, gold purchases doubled in 1970 just prior to a massive devaluation of the lira; in the Yemen five years of civil war in the 1960s made gold coins the most welcome internal currency; in Indonesia the uncontrolled inflation after the fall of President Sukharno made it one of the best markets for gold in Asia for several years; while the war in South Vietnam generated an equal addiction to gold among the Vietnamese.

Such emergencies, however, have merely strengthened a continuing trend for more and more of the world's newly mined gold to pass into private hands. As we have already remarked, every year since 1959, except 1963, more gold has gone into private hands than into monetary reserves. And since 1966 the equivalent of all newly mined gold has gone into private hands; in terms of tonnage monetary reserves of gold are now lower than they were at the end of 1965. Where does it go? Without doubt the major part is fabricated for jewellery and, to a lesser extent, industry; we shall look at that specialised field in some detail in the next chapter. But a substantial proportion remains in bars (often handy kilo, 500-gram or 100-gram sizes) for that mysterious fellow 'the hoarder'.

Europe's greatest hoarders are the French. Two world wars fought across their land in a generation coupled with sixteen devaluations of the franc since 1914—making it worth less than 1/250th of its value in that year—have fostered faith in gold. 'After being robbed for sixty years they are determined to hold something that cannot be destroyed by war or devaluation,' a former senior manager at the Bank of France once remarked to me. No one knows precisely how much gold is held privately in France, but most conservative estimates put it at 4,000 to 5,000 tons, which is about one-eighth of all non-monetary gold. This golden hoard was originally amassed in the years between the two world wars, but a good 1,500 tons has been added since 1945. The hoarding habit after the war was fostered by fears that the communists might take control and it only really slowed down with the stability that de Gaulle brought to French political and economic life after 1958. The gold crisis of March 1968, however, and the student riots in Paris a couple of months later stimulated the hoarders again. They snapped up a good 450 tons (then worth about $500 million) that year, and added 75 tons in 1969 before the franc was devalued. Once bought, the gold is rarely disgorged; it stays under the proverbial mattress. Even the higher gold prices of 1971 and 1972 tempted very little out, perhaps no more than fifty tons in each year; just enough, in fact, to keep the French jewellery industry ticking over nicely and save it the trouble of getting gold from outside France.

Most Frenchmen buy the kilo bar—the *savonnette* they call it, for it looks just like a bar of soap. Their other favourite is the 20-franc gold Napoleon; originally the two Napoleons minted some 265 million of these coins. Today that supply is supplemented by fakes from Italy and Beirut. The fact that the coin fetches a premium of anything between 45 and 60 per cent over its gold content does not seem to deter buyers.

The buying and selling of the coins and bars is funnelled through the Paris gold market, which holds its own fixing each weekday at 12.30 in the basement of the Bourse. The Bourse Committee—the Chambre Syndicale des Agents de Change and six leading French banks (Banque Nationale de Paris, Crédit Lyonnais, Société Générale, Banque de Paris et des Pays Bas, Banque de L'Indochine, and Compagnie Parisienne de Réescompte) are represented at the fixing; Compagnie Parisienne de Réescompte presides over the ceremony. Each firm has its own tiny booth in the fixing room where a couple of clerks crouch over direct telephone lines to their offices. In the centre of the room the brokers themselves stand around a waist-high oval counter, clutching little slips of white paper with their buying and selling orders. Behind them on a large electronic board the prices for bars and coins flash up as the ceremony proceeds. The fixing starts by determining a price for the 12½ kilo bar (the 400-ounce good delivery bar), then moves on to quote for the kilo bar, the 20-franc Napoleon, the U.S. $20 double eagle and even the Mexican 50-pesos coin. A broker who is selling will shout 'Je l'ai', while buyers counter with 'Je le prends.' To the outsider it proceeds with bewildering speed, as the brokers quickly jump in with bids as the messengers from their booths nearby come dashing up with new instructions. On a normal day at least 500 kilos changes hands; about 200 kilos of it for the jewellery industry, the rest is the hoarders playing the market.

Despite its widespread reputation the Paris market is essentially a domestic one. The import and export of gold from France for private buyers has been tightly controlled, except for a short period in 1968 when exchange controls were relaxed. In theory, therefore, the market exists in a closed environment, just recirculating the gold within the country. In fact, there has never been the slightest difficulty in smuggling gold into France, mainly from Geneva. Indeed until 1968, upwards of a 100 tons entered France clandestinely each year to satisfy the hoarding demand. The gold was paid for in French banknotes smuggled out to Switzerland and exchanged there for Swiss francs, which, in turn, were used to buy gold. The whole operation was very well organised by two or three syndicates in Geneva, so that, despite controls, France was amply supplied with all the gold she required. There was also a nice loophole, called *repatriement*, that enabled substantial quantities of gold to come in quite legally. A Frenchman who had been living abroad was entitled to repatriate to France any gold he had purchased outside the country. An awful lot of Frenchmen turned out to have been holding gold abroad.

When all exchange controls were removed for a short while in 1968 it looked as if Paris would make a major bid to become a truly international gold market. There was even talk of her trying to handle some South African sales. However, the reimposition of exchange controls in November 1968 damped these aspirations smartly and the market reverted to its traditional role as the middleman between the local hoarders. Not that all the gold purchased each day in that crowded room below the Bourse stays in the country. Many Moroccans and Algerians working in France buy coins or kilo bars and send them home by devious means to their families in North Africa.

The great issue, now that higher gold prices are the order of the day, is will the French dishoard in large quantities? They could cause a sizeable upset on the free market if all those 5,000 tons of kilo bars suddenly emerged from the mattresses. So far there has been little sign that they will. The Frenchman, after all, has been buying gold over the years to avoid *losing* money, rather than *make* it. The sudden jump in the gold price is simply an excellent illustration of what a wise fellow he has been to buy gold. Why on earth sell it now? So although there has undoubtedly been a little profit-taking, in the main the French are sitting on their gold as staunchly as ever. The prospect for the future is that they will buy less, but not dishoard. The young generation show little interest in putting their savings into gold; but the older will hang onto their little hoard until they die (and when they do die their kilo bars will be judiciously handed on to other members of the family without any embarrassment of death duties).

Compared with the French other long-term hoarding in Europe is on a modest scale. The Austrians, Belgians, the Italians and the West Germans all indulge from time to time, but their purchases are a drop in the ocean compared with France. Investigations by Consolidated Gold Fields, for instance, have indicated that long-term hoarding (as opposed to speculative buying) in West Germany was only 15 tons in 1968 and perhaps 3–5 tons a year since then. The strength of the German economy has obviously helped to eliminate the feeling for gold which persisted immediately after the war; today the German prefers to spend his money

on consumer goods, interest earning assets or holidays by
the Mediterranean.

Much of the gold buying that does continue in West
Germany and other European countries is in coin—the
sovereign, the Belgian 20 francs Leopold II, the Swiss
20 francs, the U.S. $20 double eagle and the $10 eagle
liberty. The persistent best-sellers, however, are what are
known as Austrian 'trade' coins. These are restrikes of the
1 and 4 ducat, the 10, 20 and 100 crown and the 4 and
8 franc pieces of the old Austro-Hungarian empire. The
ducats, dated 1915, are the most popular, and have the
distinction of being the purest gold coins minted anywhere
—they are 986 and 1/9 fine gold (23 and 2/3 carat). The
coins are made at the Austrian Mint in Vienna on behalf of
four Austrian commercial banks, who work closely with
Mocatta and Goldsmid in London and the Swiss banks.
More than 20 tons of gold goes into these 'trade' coins
every year. Many of them are sold within Austria to the
local people or tourists, others are distributed widely
throughout Western Europe. A substantial proportion,
however, filter into Eastern Europe and Yugoslavia, where
they sell well on the black market. The coins are produced
in such vast quantities that they carry very little premium
(normally around 7 per cent) over their gold content. They
thus represent excellent gold value for money and appeal
to a wide range of hoarders. By comparison, such coins as
the U.S. $20, on which the premium over the gold content
is usually around 70 per cent, or on the German 20 marks,
on which it fluctuates from 140–190 per cent, seem
expensive.

However, the 'cheap' Austrian coins now have a rival in
the South African Krugerrand. This new coin, minted since
November 1970 by the South African Mint, contains
precisely one troy ounce of gold with a fineness of 916·6
(i.e. 22 carat). It is a legal coin in South Africa and was

conceived by the Chamber of Mines there with the idea of enabling the ordinary person to hold gold in money form. When the idea was first mooted in 1969 many European coin dealers showed considerable scepticism, but the South Africans persisted. And the coin has been a great success— although perhaps not in quite the way the South Africans intended. In 1970 and 1971 they minted over three-quarters of a million Krugerrands. The coins sold in modest amounts in Belgium, Canada, Italy and Britain; but in West Germany, to everyone's initial surprise, they became the hot favourite. At least 80 per cent of all the coins were sold there. Why were they so popular? Well, not because the Germans were anxious to bury them in the cellar. The coins were exempt from value added tax, and were thus a much cheaper way of buying gold for jewellery or industry than purchasing gold alloys on which VAT was charged. Sad to relate, many of those fine Krugerrands have simply been melted down for their gold.

Avoiding VAT and other kinds of taxation is one of the prime motives for gold buying by individuals or small family business in Europe these days. Profits put into gold may not earn interest, but they are not liable to taxes or death duties either. 'Dodging taxation is an increasing reason for holding gold,' a Swiss gold dealer pointed out, 'After all, if a man dies with a couple of kilo bars in his safe, who is to know about them? His son just removes them. But his shares, property, even his paintings are all known and are liable to death duties.' And what would the dealer recommend as a sound gold portfolio for the small buyer? 'It's good to have 10 per cent or 15 per cent of your money in gold,' he advised, 'which should be half in bars, half in coin. Among the bars you might have twenty 100-gram bars, as these are very easily disposed of (each is worth about $285 with gold at $90 an ounce). The rest can be in kilo bars. Then for coins, take some Austrian ducats or South

African Kruggerrand on which there is little premium over the gold content, together with plenty of gold coins of your own country.'

The most substantial gold buying to dodge taxes and exchange controls has usually been in Italy. Both industrial users and hoarders have long preferred to buy their gold in Chiasso on the Swiss-Italian border and smuggle it in. The jewellery manufacturer, for instance, was thus able to hide the true size of his business from the taxman. While anyone who wanted to get lire out of Italy to exchange them for a more desirable currency like Swiss francs simply bought a bar of gold on the Italian black market, smuggled it out again to Switzerland and sold it for Swiss francs, which he deposited in a numbered account. There is quite a regular cycle of gold flowing into and out of Italy for this little dodge.

The only people in Western Europe who have no tradition of sheltering under such a golden umbrella are the British, the Dutch and the Scandinavians, which is quite in keeping with their national character. They tend not to be nations of tax dodgers, nor have they suffered the experience of their currency becoming worthless overnight (despite all the problems of sterling nowadays). The British, of course, are not permitted to hold gold privately, although since 1 April 1971 they have been allowed to buy and sell gold coins. However, this recent relaxation on gold holding has not triggered off any major rush into gold coin. Nevertheless, two members of the London market, Johnson Matthey and Sharps, Pixley, have found it worthwhile to expand their coin departments to cater for any British coin hoarder who comes by. 'There was a pent up demand for coins,' one dealer told me, 'and we have sold thousands of sovereigns and Krugerrands.' Most of the purchasers, however, seem to be city executives buying a handful of sovereigns to give their nephews and nieces as Easter or Christmas presents. That kind of demand is hardly likely to start a gold rush.

Indeed, the Royal Mint, which makes the Queen Elizabeth sovereigns, has been selling from stocks since 1968. In the late 1950s and early 60s they did mint substantial quantities of sovereigns each year, all for export. Most of them went to Greece, where the sovereign circulated so widely that it was almost a secondary means of exchange beside the drachma; all house prices, for example, were quoted in sovereigns. But the Bank of Greece tightened up on the sovereign traffic in 1965, in an effort to check the drain on their foreign exchange caused by the sovereign imports. All private trading in sovereigns was forbidden, and people were permitted to sell sovereigns only to authorised commercial banks. Although a black market in the coins persisted, the major inflow ceased (and the Royal Mint lost its best customer). After a bumper year in 1968 when the Royal Mint used over 26 tons of gold for sovereigns, the existing stocks have met all the requirements of the international (and new British) market since then.

The 'traditional' hoarder, with his little stock of coins and bars, is, if anything a dying force in the gold markets of Europe nowadays. There is, however, another category of gold buyer whose impact on the market over the last decade has been substantial. This is the gold investor, or, as Paul Jeanty of Samuel Montagu dubbed him at *The Financial Times* conference on gold in October 1972, 'the offshore hoarder'. He is a much more sophisticated fellow than the traditional peasant hoarder, although, like him, he may have some inbred feeling that lures him to gold. This offshore hoarder, according to Mr. Jeanty, 'holds his gold, not necessarily in his own name, in a bank abroad, usually a Swiss bank—but it can just as easily be in Canada, London or another European centre, or again, he may simply buy a gold certificate.'

The offshore hoarder may be an individual such as an Arab oil sheikh, an African or South American politician or a Texas millionaire, but he is just as likely these days to be the accountant of a multi-national company who advises his directors they should keep a small part of their reserves in gold. The difference between these buyers and our traditional hoarder is that they do not normally take delivery of the gold; it is held for them instead by bullion dealers in Switzerland, and, to a large extent, in London. The gold, therefore, is never moved physically; it sits in the dealer's vaults and the client may never actually take delivery at all. The higher gold prices after June 1972 and the dollar devaluation of February 1973, swelled the band of hoarders. They snapped up at least 100 tons in 1972 and accounted for more than 25 per cent of all purchases in February and March 1973.

Each buyer specifies whether he wants his gold 'allocated' or 'unallocated'. If allocated, he will be given a document listing the exact weight, number and assay of each bar he owns. He can come in at any moment and claim them or, if he so wishes, stand and gloat over them as they rest in the vaults alongside the vintage wines and port for the directors' private dining room above. For this privilege he pays a fee of at least 2 mille ($2,000 on $1 million in gold) for storage and insurance. Most gold buyers opt instead for unallocated gold, which means that specific bars are not assigned to them. They can, however, claim gold at any moment from dealer's stocks (though in practice it is highly unlikely that they will ever take physical delivery). The bullion dealers make no charge for unallocated gold and encourage their clients to purchase their gold in this form. It does, after all, provide the bullion dealer with a much larger working stock, provided his commitments are covered. Perhaps if a bullion dealer went broke, the owner of unallocated gold would be at a disadvantage; the allocated gold would obviously auto-matically revert to its true owner, but the unallocated gold would be bracketed with the remaining assets of the dealer.

The gold investor can, if he prefers, simply buy gold certificates. The most successful gold certificate scheme was launched by Samuel Montagu, the London bullion dealers, together with the Bank of Nova Scotia in Canada, Union Acceptances Ltd. in South Africa (the merchant bank subsidiary of Anglo American) and Deutsche Bank A.G. in West Germany in the late 1950s. Their golden gold certificate, measuring $8\frac{1}{2}$ by $11\frac{1}{2}$ inches, boldly declares that 'Samuel Montagu & Co. Ltd. undertakes to deliver . . . fine ounces troy of gold to . . . upon surrender of this certificate at its office in London, England.'

Actually there is really little difference between gold certificates and unallocated gold, but perhaps a flamboyant certificate gives the owner a comforting feeling. The real issue is whether he wants to display his holding or keep quiet about it. Most gold buyers prefer to be exceptionally discreet.

An African or South American ruler, for instance, may be overthrown any night by a *coup d'état* after supper; he can hardly nip down to the bank to draw out some money or buy travellers cheques before he flees over the border under cover of darkness. But a few kilos of gold held anonymously in a Swiss vault can provide him with more than bread and butter in exile. This specialised club of hoarders has included in the past such men as Juan Peron of Argentina, Rafael Trujillo of the Dominican Republic and King Farouk of Egypt.

The oil-glutted Arab sheikh is also, by repute a great gold buyer. No doubt to begin with many of the rulers in the Middle East did invest all their sudden riches in gold, but over the years they have become much more sophisticated. While they may keep a small nest-egg in gold for ever, many of them have liquidated their main gold holdings long ago and put their money instead into more profitable property, stock market, or currency dealings in Europe. The high interest rates available on the Euro-dollar market in 1969 and 1970, for instance, wooed many Middle East potentates away from gold. Nevertheless they are always ready to shift back to it at the drop of an exchange rate. During the currency uncertainties of 1972, Kuwait officially asked the oil companies who take her oil to please pay in gold.

The major gold investors, however, are not dictators hedging against a coup or even Arab sheikhs; they are businessmen and companies in Western Europe who like to keep a small slice of their assets in gold. 'I can think of a hundred people, each worth $100 million who like to keep a little in gold,' a London bullion dealer told me some years ago. And they are joined by industrial concerns, small private banks, insurance companies and investment trusts all of whom draw comfort from a permanent golden umbrella ready against the storm clouds of devaluations or restrictions on remittances in times of currency crises or political upheavals. Investment buying was particularly strong in the late 1950s and early 1960s, but slackened off by 1966. In the late autumn of 1966 one Zürich bullion dealer told me, 'The buying of gold for investment accounts here has diminished of late, because monetary tightness has hit even the big firms, who were putting a considerable amount of their liquid assets into gold five or six years ago.' However, although they may not have bought more gold, they did not necessarily sell their existing nest-egg.

The gold rush of 1968, of course, brought them back in force, but that was much more of a speculative fling. Most firms who bought in that stampede liquidated their positions over the next couple of years. The investment buyer, now re-christened 'off-shore hoarder', only really

re-appeared in 1972 and 1973, purchasing gold chiefly in
lots of perhaps 100 or 200 kilos at a time.

There is always great debate on how much of this
investment buying represents Americans acting discreetly
through third parties in Europe. The Americans, of course,
have been forbidden to hold gold privately in the United
States ever since 1934. They were permitted to buy overseas
until 1961, when President Eisenhower, in one of the final
acts of his administration, banned further purchases. He
also decreed that any gold already held abroad must be
sold before 1 June 1961. Since then any gold buying has
been under the counter. The convenient gold market for
Americans is just across the border in Toronto, but it is
likely that most American buying has been through the
more anonymous cloak of Swiss banks. The United States
Secret Service does try to hunt down Americans who flout
this law, so gold buying through subsidiary companies in
Europe is the safest policy. Frankly, it is impossible to gauge
the true volume of American illicit buying of gold. Under
normal circumstances it is certainly fairly small, but in a
crisis, such as March 1968, it can be substantial. Indeed, in
many ways, it was the sheer volume of American private
buying in that gold rush that made it impossible for the international gold pool to keep supplying the London market.
The purchasers ranged from private individuals to large
American corporations who were active in Europe; even
some American banks bought heavily through 'correspondents' in European banks. Most got away with it.
However, the U.S. government did later bring charges
against Bernard Cornfeld and his famous Investors Overseas
Services (IOS) for alleged illicit gold buying at that time. A
U.S. attorney alleged in a New York Federal District Court
in December 1971 that an IOS subsidiary, IIT purchased
$20 million in gold on 8 March 1968 and a further $10
million four days later. Cornfeld himself was alleged to have

personally caused another IOS unit, Fund of Funds, to acquire $6·27 million in gold.

The Americans were offered another backdoor to gold buying in November 1972 when the Winnipeg Commodities Exchange just across the border in Canada started a market in gold futures. Officially Americans cannot even dabble in gold futures. 'We have determined that the purchase of a gold futures contract is the same as a purchase of gold itself,' Thomas W. Wolfe, director of the Office of Gold and Silver Operations in Washington D.C., sternly told reporters. Only U.S. firms licensed to trade in gold were permitted to participate, but a futures contract had to count as part of their authorised stock. The Canadians, however, were confident that this ruling would not stop U.S. buyers dealing through third parties in Canada. Indeed, the whole success of the Winnipeg experiment really depended on the extent of American participation.

The lobbying for America to relax its regulations on the private holding of gold is intense. The campaigners argue that it is quite inconsistent for the United States government to downplay the role of gold as a monetary metal and forecast its demise from the monetary system, while at the same time refusing to allow its citizens to purchase it like any other commodity. No doubt if the regulations were relaxed there would be substantial private buying by Americans—but primarily for speculation rather than long-term hoarding. The highly volatile gold price is now a great attraction to the speculator, whether from Europe or the United States, who can now notch up a handsome profit in an afternoon with price changes of $1 or more per ounce in an hour or two. Up to 1968, when the gold price rarely moved outside the narrow confines of $35·00 to $35·20, there was not much to excite the speculator. The soaring prices since early 1971, however, have made gold, as opposed to old-fashioned hoarding or even long-term investing, a tempting prospect for the knowledgable speculator. In the summer of 1972, when the price had gone through the $60 barrier with ease, the market was dominated by speculators out for a quick turn, while the long-standing gold markets of the Middle East and Far East were at a complete standstill. As Walter Frey of the Swiss Bank Corporation put it, 'the market had become a mere playground for speculators.'[1]

The prospect for the 1970s, therefore, is that although the old-fashioned hoarder may opt to spend his spare cash on colour television, the gold investor and the speculator will be very active. After all the man who bought 2 tons of unallocated gold in 1970 for about $2·2 million, could have sold it in April 1973 for over $5·5 million. For the first time in almost forty years gold has paid a dividend.

[1] *The Financial Times* conference on Gold, October 1972.

The Industrial Arts

A brain-teasing question for a television quiz programme would be 'What do a bottle of after-shave lotion, the Concorde supersonic airliner, an inn at Bloomfield Hills, Michigan, the moon buggies of Apollo astronauts, and an insurance company building in Sydney, Australia have in common?' The answer to this conundrum is perfectly simple—gold. Liquid gold, like molten sunshine, is sprayed on the after-shave lotion bottle to give it more man-appeal; the exhausts of the Rolls-Royce engines for the Concorde are gold plated to reflect heat; the Fox and Hounds Inn at Bloomfield has a roof of gold-plated tiles off which the summer sun bounces; gold foil shrouds the miniature television camera and other sensitive instruments on moon buggies to protect them from the fierce rays of the sun in space; while the Australian Mutual Provident Society building in Sydney has glass walls shot through with finely spun gold to deflect light and heat, thus saving air-conditioning bills.

The new uses of gold in science and industry multiply every year, stimulated by its unique properties of resistance to corrosion, its excellence as an electrical conductor, its reflectivity and its incredible malleability (it can be beaten into a foil less than five-millionths of an inch thick). Its virtual indestructibility frequently gives it superiority over any other metal. 'When people want something to be 500 per cent reliable for twenty years, we recommend gold,' says a scientist at Johnson Matthey, the London precious metal experts.

This nobility has appealed to goldsmiths for over 5,000 years. 'Gold is the Child of Zeus,' wrote Pinder, 'neither moth nor rust devoureth it.' When archaeologist Howard Carter discovered the tomb of Tutankhamun in Egypt in 1922, the golden treasures of the boy king, who reigned from 1361 to 1352 B.C., were perfectly preserved. The king's body was encased in a coffin of solid gold nearly 2 millimetres thick, weighing 242 pounds; the head of the mummy itself was shrouded in a great mask of beaten gold. The golden throne nearby was adorned with delicately worked scenes showing the young king being anointed by his queen. These treasures, now in Cairo Museum, look as perfect as when they were first created 3,000 years ago. Their extra-ordinary magic drew almost a million people to the British Museum in 1972, when fifty of the finest items were displayed in a special exhibition mounted by *The Times* and *The Sunday Times*. Whole families queued patiently for hours outside the museum for a glimpse of these supreme examples of the goldsmith's art.

Benvenuto Cellini's solid gold salt cellar, designed for Francis I of France and now in Vienna's Kunsthistorische Museum, is equally untarnished by age.

'It was oval in form, standing about two-thirds of a cubit,' wrote Cellini in his egotistical autobiography, 'wrought of solid gold, and worked entirely with the chisel. It represented Sea and Earth, seated, with their legs interlaced, as we observe in the case of firths and promon-tories. The sea carried a trident in his right hand, and in his left I put a ship of delicate workmanship to hold the salt . . . I had portrayed Earth under the form of a very handsome woman, holding her horn of plenty, entirely nude like the male figure; in her left hand I place a little temple of Ionic architecture, most delicately wrought, which was meant to contain the pepper.' It is hardly surprising that when Cellini first unveiled his masterpiece to

Francis I, 'He uttered a loud cry of astonishment, and could not satiate his eyes with gazing at it.'[1]

Goldsmiths rarely find such generous or appreciative patrons today. Most gold goes into such plebeian items as wedding rings, fountain pens, teeth fillings and the gold watches handed out endlessly by company managements to loyal workers after twenty-five years' service.

Although few people nowadays commission such exotic articles as solid gold salt cellars, many millions can buy gold jewellery for the first time. Gold, after all, did not rise in price from 1934 until 1968, while average earnings soared. Gold ornaments, once afforded only by the rich, are now within the means of most men—and women—in the developed countries.

'Gold is cheap in terms of work,' remarked a Swiss banker, gazing reflectively out of the window of his office in Basel. 'Take a tram driver here. On his wages he can buy his wife a gold wedding ring and even a gold bracelet once in a while.'

The demand for gold moves forward in harness with the wealth of a nation. The Americans, for example, buy more gold jewellery per head than the Indians—for all India's reputation as a sponge for the metal; in 1971 over 125 tons of new gold was fabricated into jewellery in the United States for her 200 million people, compared with about 200 tons on the Indian sub-continent for almost 750 million people. While Western Europe, with a population of about 350 million, in the same year used 420 tons of gold in jewellery. Furthermore, world-wide investigations by the two major mining finance houses, Charter Consolidated (the London associate of Anglo American, the largest producers of gold) and Consolidated Gold Fields have both revealed that the demand for gold for fabrication in jewellery,

[1] *The Life of Benvenuto Cellini,* written by himself, Phaidon Press Ltd., London, 1949.

dentistry, coins and industry was running almost level or slightly ahead of the production of newly mined gold in the non-communist world each year from 1968 to 1971.[1] Gold bought for long-term hoarding or quick speculation comes on top of that. The Consolidated Gold Fields calculations indicate that in 1971 some 1,412 tons of gold was fabricated; this included 1,058 tons which went into jewellery, 76 tons into teeth, 91 tons into electronics, 75 tons into other industrial and decorative uses and 111 tons into coins, medals and medallions. Gold Fields also suggest that, depending on the price, the demand could rise to about 1,625 tons by 1975 and to over 2,000 tons annually by 1980. By comparison, the production of gold in non-communist countries is running at between 1,200 and 1,300 tons a year and is declining. The higher prices of 1972 did lead to a drop in fabrication to around 1,200 tons, but this was chiefly in developing countries. American, West European and Japanese use was not affected.

While it is difficult to be precise in the gold business, where so many bullion dealers, smugglers and jewellery manufacturers decline to reveal their exact turnovers, the mining finance houses research has pinpointed more closely than anyone had ever attempted before the true extent of gold usage in 'the industrial arts'. Inevitably, there are errors in assessing consumption in individual countries, but the overall message is clear: gold is ready to stand on its feet as a commercial, non-monetary metal, just like silver, or copper or lead.

The Charter Consolidated and Consolidated Gold Fields findings were initially greeted with some scepticism both in Europe and even on home ground in South Africa. Old hard line advocates of gold's intrinsic value as a monetary metal

[1] *Gold: A World-Wide survey*, Charter Consolidated, London, 1969 David Lloyd-Jacob and Peter Fells, *Gold 1969, Gold 1971, Gold 1972*, Consolidated Gold Fields, London.

were clearly put out. As a senior manager of Anglo American in Johannesburg put it, 'They felt that to admit a non-monetary role for gold would weaken its monetary role, and that such downgrading could be dangerous.' The fact is, however, that nowadays jewellery can absorb each year all the gold South Africa producers.

Since pure gold is too soft and liable to wear to be practical in most jewellery, it has always—even in ancient times—been alloyed with other metals. The amount of gold in an article of jewellery is defined by the 'carat' scale. (The word 'carat' comes from the Italian *carato*, the Arabic *qirat* and the Greek *keration*, all meaning the fruit of the carob tree. The horn-like pods of this tree contain seeds which were once used to balance the scales in Oriental bazaars). Pure gold is 24 carat. The proportion of gold in jewellery varies considerably from country to country, depending on whether there is some hoarding motive in buying the jewellery, or whether it is bought purely for adornment. In the *souks* of Morocco, Beirut, Kuwait or the markets of Istanbul, Bombay, or Bangkok most gold jewellery is (or purports to be) 21 or 22 carat, because the ordinary people there are buying the gold at least partly as a form of saving or hedge against inflation. In Western Europe, however, jewellery is usually 14 or 18 carat. In France and Sweden the lowest legal limit is 18 carat (anything below is regarded as costume jewellery); in Spain virtually all jewellery for the local markets is 18 carat; Belgium, Denmark, Luxemburg, the Netherlands, Norway and Switzerland do not recognise as carat jewellery anything stamped with a rating of less than 14 carat. In Britain the lowest standard accepted is 9 carat, while Italy and West Germany permit 8 carat. Most gold jewellery in the United States is 10 carat, although there has been a rising trend in favour of 14 and even 18 carat in recent years.

The advantage of lower caratage is that the colour of the

gold can range through white, green, yellow and red hues, depending on the balance of other metals (usually silver or palladium) with which it is alloyed. White shades, for example, are achieved by alloying the gold with silver and nickel or palladium; red gold contains some copper; greenish tints are produced by varying proportions of copper and silver, nickel or palladium.

In many countries, not only in Western Europe, but even in the Lebanon, the gold jewellery industry is policed by some form of hall-marking system administered by a central office. France's *Bureau de la Garantie*, for instance, impose strict controls on manufacturers as does the Dutch Assay Office. The tightest regulation is in Britain where assay offices in London, Birmingham, Sheffield and Edinburgh maintain careful surveillance of all carat gold production. Their hallmarks—an uncrowned leopard's head for London, an anchor for Birmingham, a rose for Sheffield and a castle for Edinburgh—are the guarantee that the article has been assayed and is indeed 9 or 14 carat. The Americans, by contrast, allow more leeway; the law permits a tolerance of half a carat either way. A loophole that the industry takes full advantage of, so that a bracelet stamped 14 carat will almost inevitably be $13\frac{1}{2}$ carat. 'We aim to make all our 14 carat gold exactly $13\frac{1}{2}$ carat' a major American supplier of gold alloys admitted.

For generations gold jewellery manufacture has been a family affair. Even today in most countries jewellery manufacture tends to be in the hands of hundreds of small firms employing ten or fifteen people. Often they are all gathered together in one or two towns which have a long history of craftsmanship in jewellery. In the United States, for example, where almost 1,500 firms are involved in gold jewellery, the focal communities are Attleboro, Massachusetts and Providence, Rhode Island. The town of Pforzheim, just west of Stuttgart, is the home of over 400 of West

Germany's 900 jewellery manufacturers and is responsible for 70 per cent of their output. While in Italy over 10,000 of the 20,000 people living around Valenza, a small town between Milan and Genoa, work for 1,300 firms there.

Although the goldsmiths' traditional craft lives on in these communities, the inevitable trend is towards mechanisation. The craftsmen, working slowly by hand with a few simple tools and a leather apron spread between his knees to catch the precious filings of gold, is on his way out. Major semi-fabricators like Johnson Matthey in Britain, and Engelhard and Handy and Harman in the United States, and Tanaka in Japan supply the individual manufacturers with carat gold alloys in all shapes and sizes—long rolls of wire, wafer thin sheets and slender tubes. So wedding rings can be snipped off a 6-foot long golden tube and fashioned swiftly on a lathe. Machines spew out gold chain by the mile; one machine uses over a kilo of gold a day, which would keep most craftsmen going for six months.

The world's largest manufacturing gold jewellers are Gori and Zucchi whose modern factory in the Tuscan hill town of Arezzo, near Florence alone uses more gold in a year, than the entire jewellery industry in Britain, France or Switzerland. The business, built up since World War I by the marketing flair of Mr. Gori and the technical skill of Mr. Zucchi, has thrived on the mass production of rings, chains, coins and medals. Gori and Zucchi have their own refining and semi-fabricating plants, while their production control is supervised by an IBM 360 computer. A far cry, indeed, from most other little jewellery workshops around the world, but immensely successful.

Gori and Zucchi, together with other large manufacturers like Fibo and Donnagemma in Vincenza and Balestra in Bassano del Grappa, have made Italy the world's leading producer of cheap, mass-produced jewellery. The jewellery industry there now uses up to 200 tons of gold a year, one-sixth of non-communist world production. Most of it is for export. Italy's largest customers are West Germany—indeed many German firms buy their gold in Switzerland, have it delivered direct to an Italian manufacturer for fabrication and then simply import the finished article. But Italian jewellery has been carrying all before it in markets everywhere from Beirut and Kuwait, to Hong Kong and Rio de Janiero. Walk into almost any jewellers in these cities, where fifteen years ago all jewellery would have been made in some little *artelier* over the shop, and ask to see a nice selection of 18-carat rings or bracelets. 'Are they locally made?' 'No, no, from Italy, they make jewellery much more cheaply than we can.' Furthermore the Italian designs are more sophisticated, even though the jewellery is mass produced. Gori and Zucchi, for instance, have 30 designers on their staff keeping them abreast of the latest trends.

Although no one can really rival the Italians, there is an enormous market in the United States for mass produced 'class rings' for schools and colleges. These are manufactured by a handful of large manufacturers such as Josten's of Owatonna, Minnesota, L. G. Balfour in Attleboro, Massachusetts and Dieges and Clust Co. in Providence. The rings themselves are quite simple 10-carat pieces (meaning they are actually $9\frac{1}{2}$ carat, because of the tolerance permitted in the United States), but they sell by the hundred thousand. Indeed, almost one third of all the gold used in carat jewellery in the United States goes into class rings—nearly 40 tons, for example, in 1971. By comparison wedding and engagement rings in the United States required only about 30 tons of gold that year.

Amidst this plethora of mass-produced jewellery, the true goldsmith still gets some encouragement these days to display his art. Although the days of the $250,000 gold dinner services for Indian maharajahs are over, there are still such exotic little delicacies as 18-carat gold vanity cases carved like the bark of a tree or Easter eggs in gold to be picked up for a few thousand dollars at Tiffany's in New York, Collingwood's and Garrard's in London, Cartier and Boucheron in Paris and Bulgari's in Rome. Moreover, a new generation of creative goldsmiths like Andrew Grima and John Donald have achieved great success in Britain by opening their own shops. John Donald has his strategically placed on Cheapside in the City (which was Goldsmith's Row in the Middle Ages), while Andrew Grima's dominates a corner in Jermyn Street, that elite little shopping byway parallel with Piccadilly. And nearby Collingwood, an old family jewellery business on Conduit Street, have adopted another young designer Stuart Devlin as their 'named' goldsmith.

The importance of this new trend is that it is giving much more encouragement to young designers to work in gold; previously their work was sold anonymously and they gained none of the glory. 'The young designers have made a definite chink in the armour of tradition,' according to Graham Hughes, art director of the Goldsmith's Company, 'and the designing of modern jewellery has become big business.'

The skills of the modern goldsmith were never better displayed than in the remarkable exhibition at Goldsmith's Hall in London in 1971 of the work of Louis Osman. The pride of the exhibition was the elegant crown Osman made for the investiture of Prince Charles at Caernarvon Castle in 1969. But in many ways such lavish pieces as a golden cream jug, a gold tea caddy and a tempting gold wine cup that cried out to be filled with a fine 1961 first growth claret were more intriguing symbols of an affluent society.

Nowhere, however, has gold become such a symbol of affluence as in Japan. While the businessman in Britain may give his clients a bottle (or even a case) of scotch at Christmas, the Japanese are growing increasingly fond of dispensing 24-carat gold presents as company *billet doux* to favoured customers. At Yamazaki's or the jewellery section of the great Mitsukoshi department store in the centre of Tokyo, the visitor is faced with an extraordinary display of solid gold teapots, tea caddies, vases, chopsticks and even wall plaques with a charming scene of ducks in flight. The great teapots, each containing nearly two kilos of pure gold, sell for the nice round sum of 3 million yen (just under $10,000). The vases, decorated with cherry blossom designs, are even more expensive. Variations on the golden theme seem endless. Yamazaki can offer a fully rigged sailing ship in 18-carat gold, laden with barrels of rice and butts of saki (all in gold), or a solid gold rickshaw. While in 1972, which was the 'year of the rat' in the Far East, shelves were laden with little gold rats of all shapes and sizes to suit all purses.

The Japanese have also become among the most enthusiastic collectors anywhere of medallions struck to mark special occasions. The craze really started there when the Japanese mint cast a series of medals to commemorate the 1964 Olympics. They were snapped up like hot cakes and soon commanded four times their original price. A host of private mints then jumped on the bandwagon with medallions for all seasons. The Winter Olympics at Sappho, the Emperor's visit to Europe, even a Lions International Convention in Tokyo all triggered off a new flood of medals. Many businesses have got into the act by commanding gold medals to be struck to mark the opening of their new factories or the new model of a car.

The medallion craze is not limited to Japan. Mexico did splendidly with gold medals to herald the Olympics in 1968 and World Cup Soccer in 1970, while West Germany cashed in on the 1972 Olympics. Although the medals are primarily for collectors rather than speculators or hoarders (because they cost initially far more than their gold content) many of them have turned out to be an excellent investment.

By limiting an edition, the resale price can soon soar if the medal proves popular, while the strong rise in the gold price in 1972 inevitably pushed their value even higher.

While the medals do not devour vast quantities of gold (perhaps 12–15 tons annually) they are all part of the burgeoning uses of gold.

The most visually spectacular new application for gold is for the gold-coated buildings, which are sprouting all across America, Australia and even South Africa.

They may lack the style and sheer richness of the great Inca Temple of the Sun at Cuzco from which the Spanish conquistador Francisco Pizarro and his men ripped seven hundred plates of pure gold, but the golden buildings have an essentially practical aim—to reflect the sun in hot climates and save air conditioning bills.

Randall Miller, the ebullient sales manager (even his visiting card is gold coloured) of the Hanovia Liquid Gold Division of Englehard Industries in the United States, admits that when he first approached architects and company chairmen with the idea of gold-encased buildings it sounded like a gimmick—and a very expensive one at that. 'But,' he says, 'the funny thing is it really works. We can offer golden roofs or walls at 50 cents a square foot.' The original 20th century gold-plated building was the Richfield Oil Building, put up in Los Angeles in 1929. It has a ceramic veneer gold surface using $12,000 worth of gold. Amid all the smog of Los Angeles the gold has not tarnished over the years. At first the gold surfaces were washed every five years with Castile soap. Since 1959 they have been cleaned once a month with a very mild solution of an all purpose cleaner called Steamite. The depression years when everyone was selling their jewellery and companies could not afford new gold headquarters, meant that the Richfield Corporation had no imitators until after World War II. But in the last few years gold has been used to adorn hotels in California, banks in Los Angeles, Beverly Hills and Dallas, and a chain of Golden Key coffee shops—all with golden roofs. Even the exterior of an Armenian church in Detroit and a Roman Catholic church in Washington, D.C., have been covered with gold. The 22-carat gold is, of course, spread thinner than butter on bread over tiles, bricks or other metal, so that a few ounces of gold can coat a modest building.

The latest notion is that gold can be thinned down even more, so that it becomes translucent. It is then applied to glass and can be used for windows. People looking out through the glass will see the world with a slight greenish hue, but the thin film of gold will be enough to reflect heat and eliminate glare from the sun. Depending on the climate, gold walls and windows could cut out the need for air conditioning entirely or reduce the size and running costs of an air conditioning plant. There is one climate for which it is not recommended—where there are frequent hail storms, for the soft gold will pit too easily.

Liquid gold in the form of an oil solution, which could be painted on glass and ceramics with a camel-hair brush, was first developed in Europe as far back as 1830. Today, at minimal cost, a film of gold four-millionths of an inch thick can be sprayed or stamped onto almost anything and after firing will have a metallic brightness. 'We can make up a real witches' brew with gold,' says G. E. Fitzgerald, vice-president of Hanovia Liquid Gold. A bottle of whisky or of cosmetics can be adorned with gold lettering or designs for less than 1 cent. Even pickle and peanut butter jars have their dab of gold added by promoters and advertising men, who feel that gold glamourises even the most mundane of products.

Technology is opening up the widest new horizons for gold in microelectronics, missiles and space vehicle development. Here the double pressure of the need for supreme reliability, coupled with the miniaturisation of equipment, gives gold the edge in many situations. Space research, in particular, calls for this implicit faith in equipment. The American Gemini and Apollo spacecraft have used gold as a shield against heat and thermal radiation inside the spacecraft and on the umbilical cords that link the astronauts to the spacecraft as they make their perilous walks into space. Silver, for example, would be hopeless in the same situation. 'It might tarnish before they got into orbit,' says a scientist. 'There is no such thing as a piece of virgin silver because there is no pure atmosphere, but gold never oxidizes and it's more likely to stay shining bright forever.'

A thin film of gold reflecting intense heat can save a tremendous dead weight that would have to be carried by jet aircraft or rockets if normal insulation of wads of asbestos was employed. Although this application for liquid gold was developed in the United States. it has been taken up with special enthusiasm by the British jet engine manufacturers, particularly in engines for fighter aircraft. The pilot of a fighter plane is sitting virtually right on top of his engine and the effectiveness of the heat shield is critical. Rolls-Royce have also found it invaluable in their engines for the Anglo-French supersonic Concorde. For the same reason, gold coatings are being used for the external parts of vernier rocket engines in the Apollo space programme to provide thermal control and maintain the engines within safe operating temperatures, while the spacecraft is boosted clear of the earth's atmosphere.

The chief value of gold in industry, however, is in electronics, where its excellence as a conductor and its low resistance make it ideal as a connector in miniaturised equipment that cannot easily be serviced. It is used, for instance, as an electrical contact in the tiny 'repeaters' that are planted at intervals along cable links beneath the oceans and in communications satellites orbiting in space.

Since the cables are laid and the satellites are launched at enormous cost they must work infallibly for years; an electrician cannot dive down to fix a dud repeater 10,000 feet below the surface of the Atlantic Ocean or buzz up to repair a satellite. In computers, which call for similar reliability and miniaturisation, wires are replaced by electronic circuits 'printed' in copper overlaid with liquid gold on strips of plastic so minute that they can barely be seen with the naked eye.

Gold's excellent conductivity has also led to a thin coating, only one-fifth of a millioneth of an inch thick, being interlaced with laminated glass in the windshields of jet aircraft, such as the VC-10, to provide electric heating within the glass to prevent icing.

These highly sophisticated applications of gold are limited, however, to the most technologically advanced countries. This is especially true in electronics, where the United States uses almost half of all the gold (about 51 tons out of 108 used in the non-communist world in 1972) and Japan nearly a quarter. The consumption of these two goliaths reflects the dominance of their electronics industries in both computers and colour television in which gold is used. The other major consumers in electronics are West Germany, Britain, France, and the Netherlands, again reflecting the significance of their electronic industries. The additional dark horse is the Soviet Union. Their industrial requirements for aerospace, computers, television and other electronic equipment must now be substantial. Michael Kaser of St. Anthony's College, Oxford, in his special analysis of the Russian gold scene for Consolidated Gold Fields has assessed total Soviet industrial demand for gold as rising from 21·7 tons in 1968 to 34·7 tons in 1971; the major slice of this would be for electronics. (The Russians also now seem to be releasing small quantities of newly mined gold to their jewellery industry each year. Up to

1969 Soviet jewellers had had to make do ever since the Revolution with forever re-working old gold.)[1]

Electronics, however, is never likely to overhaul jewellery as the prime consumer of gold. A little gold goes a very long way in electronics. Many thousands of 'printed circuits' or connectors can be stamped out of a single ounce. Some industrial users order as little as one-fifth of an ounce at a time; hardly enough to make any gold-miner rich. Nevertheless, in the last five years electronics has overhauled dentistry as the second largest consumer of gold. A modest 70–75 tons of gold suffices to fill the world's teeth each year. Once again the United States heads the league; over 20 tons of gold finds its way into American teeth each year. The Germans (14 tons) and the Japanese (10 tons) also appear to favour solid gold teeth. The British, on the other hand, hardly use it at all, which is either a reflection on the quality of the National Health Service for keeping the nation's teeth free from decay or its meanness in not permitting gold fillings.

While American and European dentists use sophisticated 18- or 22-carat alloys specially prepared in dental laboratories, their colleagues elsewhere often have to be more practical. They simply buy a handful of local gold coins and melt them down. Since most coins like the sovereign or the Mexican peso are 22 carat anyway they make excellent teeth stoppers.

Macabre hints are always turning up, of course, that one reason so many undertakers seem to make a good profit is that they have a nice sideline in selling gold from the teeth of the deceased. In the United States alone it might be a very handsome bonus since the value of gold going into teeth each year is at least $30 million (based on the monetary price of $42 an ounce).

Besides its value in repairing teeth, gold has other special

[1] *Gold 1971* and *Gold 1972*, Consolidated Gold Fields, London.

medical uses. It has been used in Europe since 1927 for the treatment of rheumatoid arthritis, being administered intra-muscularly as a soluble salt in cautiously increased doses to a level of 25 milligrams per week. The treatment is slow, but apparently there is usually some improvement after six weeks and maximum benefit in six months. Gold is also now being used in serious cancer cases, in the form of an injection of a colloidal suspension of radioactive gold. Solid gold barriers may also be used internally to protect patients' vital organs when X-ray or radiation treatment is being carried out.

Quacks and alchemists have also, of course, not hesitated over the centuries to endow gold with all kinds of other medicinal blessings. The mythical 'philosophers' stone', besides being capable of turning base metals into silver and gold, was also looked on as the 'elixir' to prolong life indefinitely. Even today in India some country doctors will prescribe 'golden pills' as remedies for skin disease, impotence and infertility.

The realisation over the last three or four years that the new uses of gold, combined with flourishing gold jewellery sales have brought consumption to a point where it now at least equals the output of newly mined gold has caused the gold producers to do some radical re-thinking. Previously the producers, especially in South Africa, felt that gold was something they burrowed out of the ground for central banks to re-bury in their own treasuries. The gold lobby's sole preoccupation was to persuade the central banks to pay them more by getting the monetary price up. Well, that has moved a minute $7 in almost forty years; the free market price for gold for jewellery and industry on the other hand stood in the $85-$90 range in April 1973—an increase of $50. So the lesson is getting home that the future of gold is primarily going to be as a non-monetary metal. The industry, therefore, is slowly gearing up to promote gold as a commodity for the first time. The Chamber of Mines in South Africa has launched an International Gold Corporation (Intergold) with offices in London, Geneva and Munich. Intergold's basic aim, besides boosting the gold price, is to work in close liaison with the local jewellery industries in promoting even greater sales of gold jewellery. Clearly the prototype for such an operation is De Beer's long running 'Diamonds are Forever' campaign. Intergold has been playing around with similar slogans—'Gold Endures . . . Gold Endears' and 'Nothing is as good as gold'—in some preliminary advertising flurries, but they are unlikely to embark on anything as elaborate as the full spate of all those dewy-eyed-love-by-the-shore diamond advertisements. Instead Intergold and the major gold producers on their own will concentrate on encouraging the gold jewellery manufacturers to polish up their own organisations and marketing techniques. But even without the aid of romantic advertisements, gold's prospects as a commodity look remarkably healthy.

12

The Golden Route to India

Which country was the largest single buyer of gold on the London market in both 1970 and 1971? The world's largest exporter of silver in 1968? And is regularly the third largest purchaser of watches anywhere? France, perhaps? West Germany? Japan? No. Dubai: area, 1,500 square miles of sand; population 70,000.

On the map the tiny sheikhdom of Dubai, a member of the Union of Arab Emirates, appears as no more than a speck at the southern end of the Persian Gulf. It may seem one of the last unexplored corners of the globe, and, with temperatures of 130° Fahrenheit in the shade and humidity of nearly 100 per cent for much of the year, one of the most inhospitable. From the air it is an isolated cluster of whitewashed buildings huddled around a small creek on the shore of the Gulf and surrounded by a limitless waste of sand. A long line of camels is strung out across the desert, heading out into nowhere. A few isolated tracks across the sand show that Land Rovers have braved the sun and sandstorms on unknown missions. In the bustling creek itself, long lines of sleek dhows are moored three abreast along the shore, taking aboard sacks of flour, tins of paint, sheets of corrugated iron and the occasional live goat to provide fresh meat during the coming voyage. Once in a while, to the excited shouts of a crowd of barefooted Arab boys, a dhow casts off and pulls out into the creek. Suddenly, as she moves away, there comes the purr of a high-powered 320-horsepower diesel engine and the dhow surges away at a full 15 knots, leaving a creaming wake on the creek as she lifts her bows to the first waves of the Gulf. And in the magic purr of that engine lies the clue to Dubai. No other community in the world thrives so completely on smuggling. 'Dubai,' says one proud Arab, 'is the smuggler's supermarket.'

Gold is the best-selling line. In 1970 Dubai imported almost 260 tons of gold worth over $300 million and representing a fifth of world production that year. The business was down slightly in 1971, with a higher gold price, but was still a healthy 215 tons. The gold trade is closely linked with silver. In one astonishing year, 1968, Dubai handled over 2,000 tons of silver, equivalent to the entire production of Mexico and Canada—the two leading silver mining countries. Other flourishing lines include $20 million worth of watches a year and enough textiles to clothe its entire population for life. Dubai even dealt in 40 tons of playing cards in 1971—which should have been enough to keep every gambling man east of Suez happy for a while.

There is no official record of what happens to the gold, watches or playing cards once they arrive in Dubai, but precious few of these goods remain in the sheikhdom of Shiekh Rashid Bin Said Al Maktoum, who, since 1938, has ruled Dubai from his white-walled palace, which looks like a set for a French Foreign Legion film. 'The ruler,' says an admiring diplomat in Dubai, 'has great commercial sense.'

What Sheikh Rashid was astute enough to realise years ago was that his 3-mile-long creek offers the best harbour and haven within striking distance of Iran, Pakistan and, above all, India. Since India, that traditional sponge for gold, has banned official imports of gold since 1947, Dubai is most happily placed to oblige unofficially. Of the 200 or more tons of gold smuggled to India each year, virtually all

arrives nowadays tucked away on the dhows that plough
1,200 miles across the Arabian Sea from Dubai. A million
dollars' worth of gold takes up no more room than a couple
of water jugs below decks and the round-trip voyage takes
ten days.

Since the gold bought in Dubai at the normal free
market price can be sold in India at a profit of up to 15 per
cent, the incentive is obvious. And if the cargo is spiced with
watches, which often command 100 per cent premium in
India, the return is even more lucrative.

It is hardly surprising, therefore, that Sheikh Rashid's
commercial acumen, has transformed his little domain into
what *The Financial Times* recently described as 'the swinging
little Emirate'. While its neighbours of Kuwait and Abu
Dhabi have grown fat on oil revenues, Dubai (which had no
oil until 1969 and only modest amounts now) had achieved
fame and fortune in trade. Visitors touch down at a
spanking modern airport that is the pride of the Gulf. On
the creek itself work is well advanced on a bustling new
harbour, with 15 deep-water berths. When it is finished in
1973 it will make Dubai the biggest port in the Middle
East. Naturally it has already been named Port Rashid,
after the Ruler.

Local merchants, whose business knowledge a few years
ago was limited to trading a handful of pearls, have become
experts in the subtleties of the gold and foreign exchange
markets and are welcome visitors in the inner sanctums of
London and Swiss bullion dealers. Most carry on their
gold business, however, behind a discreet façade of selling
washing machines, refrigerators and cars, although one or
two have branched out to open their own banks.

The gold arrives in Dubai aboard regular BOAC or
Middle East Airlines flights from London, packed in wood
fibre boxes, each containing 200 or 250 canape-sized
10-tola bars (10 tolas equals 3·75 ounces). It is usually sent

on consignment by the London bullion firms or the three major Swiss banks (who probably have cornered about two-thirds of the market). Initially it is held in the strong-rooms of the many banks that have been established there in the last decade. For many years the British Bank of the Middle East was the only bank, but since 1963 the National Banks of Dubai, Dubai Bank, the Bank of Oman, the Arab Bank, the Eastern Bank and the First National City Bank of New York have opened their doors to get a piece of the action. When First National City started in 1964, they hoped rather piously (being American-owned) that they would not have to get involved in the gold trade. But they quickly discovered that for any bank in Dubai it is as important to be able to offer clients gold as it is to have deposit accounts or loan facilities. Gold is part and parcel of banking in Dubai. So First National City applied to the U.S. Treasury for a special licence to hold gold— something granted very rarely to a U.S. bank. The Dubai merchants, taking advantage of this proliferation of banks, play one off against the other in cutting the price for gold until there is often little margin left for the banks them-selves.

At this stage and, indeed, until the gold actually reaches India, everything is perfectly legal. Dubai allows the free import and export of gold. A British warship that once stopped a dhow in the Gulf and proudly radioed to base that it had caught a gold-smuggling ship, was quickly ordered to release the vessel. No offence was being committed at that stage.

When arrangements for a shipment to India are complete, the Dubai merchant leaves his washing machine or radio store and speeds in his car to the bank to take delivery of anything up to $2 million in solid gold 10-tola bars.

Initially, the money to buy the gold in Dubai comes from the merchants themselves and from a wide network of shareholders in their syndicate. Capital from Beirut, Switzerland and even London was first employed to get the gold route established. Now Dubai seems to carry much of the cost itself. Most of the merchants will accept stakes of as little as 50 tolas of gold from individual subscribers. If someone wants to have 50 tolas constantly moving along the pipeline to India, then he will be asked to pay for the costs of the first two shipments in advance, since the money from India for the first will not necessarily be received before the next shipment leaves. After that, the operation should be self-generating. Every three months or at the end of the season the Dubai smuggler reports to his clients on how their investment has fared. In a good season they should make 8 to 9 per cent profit per voyage. Much depends on whether or not a shipment is seized by the Indian authorities. There are frequently uncertain faces in Dubai when word comes through that the Indian customs have made a haul, and each merchant is waiting to learn whose shipment has been lost. Cocktail party chitchat on the merchants and bankers circuit in Dubai keeps everyone abreast of the latest gossip. 'Old so and so looks cheerful tonight. He had a cable today from Pakistan saying they've managed to pick up that 40,000 tolas his boat had to throw overboard last week.'

To help spread the cost of losses, any large investments are distributed over several voyages. If someone stakes $100,000 this will be split over at least three voyages, so that not more than one-third is likely to be lost. And a second shipment is never sent until word is received of the safe arrival of the first. An investor has, of course, no way of checking whether or not it was his gold that made up part of a seized shipment. He has to take the Dubai merchant's word for it, and pray he is not being cheated.

The Indian client for the gold has to pay up within seven days. He does this in a variety of roundabout ways.

In the early days on the Gulf, payment was made in Indian rupee notes, which the local rulers accepted as legal tender. But as India became perturbed at this unofficial drain on her currency to pay for gold, pressure was brought to bear on the rulers to change their currency. A different coloured Gulf rupee note was therefore issued and the sheikh declared the Indian note was no longer legal tender. That foxed the smugglers only for a matter of days. The Indian 1-rupee, half-anna and quarter-anna coins were still legal tender. Soon the dhows were coming back up the Dubai creek from India laden to the gunwales with 40-gallon oil drums filled with small coins. The banks in Dubai had a major problem weighing the coins and re-packing them for shipment back to India; they even instituted a 5 per cent handling charge to disuade the smugglers from paying for gold with millions of small coins. Finally, early in 1966, the small coins were also declared not to be legal tender, and the smugglers had to think again. Now India pays for its gold in three main ways. First, through the export of silver, which is smuggled *out* from Bombay to Dubai and then flown quite legally and openly to London or Zürich. This reverse flow of silver does depend on the world silver price being high enough to attract the silver from Indian hoards. In 1968 when silver prices were at their peak, India disgorged enough silver through Dubai to pay for almost two-thirds of the gold smuggled in. Since then, however, the decline in the silver price has slowed the traffic; less than a quarter of the gold trade has been financed by silver in recent years.

A second, though relatively small source of finance, is the black market in India for dollar bills, travellers cheques and even personal cheques written by tourists. While the official exchange rate is 7·20 rupees to $1, most tourists are quickly accosted by taxi drivers offering them 10 or 11 rupees for a dollar. The taxi driver in turn sells for 10·50 or

11·50; the full black market rate has been over 13 rupees to $1.

The major source of foreign exchange for gold, however, comes from Indians—and Pakistanis—working in the Gulf itself, East Africa or Britain. Many of them send part of their earnings back home. This can be done, of course, by a straightforward transfer through a bank, but that is always at the official rate of exchange. It is much more profitable to all concerned to remit the money through the unofficial exchange dealers who flourish in every community of Asians abroad. They offer much more attractive terms. Whereas the family might receive 100 rupees through official channels, they will get perhaps 125 rupees by the back door. Everyone benefits; the Asian's family gets more money, the smugglers gain access to a pool of hard currency abroad.

The dhows that speed the gold from Dubai are usually built in Pakistan, so that on their smuggling missions they are indistinguishable from the hundreds of local dhows and fishing boats putting out from the ports of India and Pakistan. The completed craft are towed back to Dubai, where their diesel engines are installed. The eight or nine crew members are always Pakistanis or Indians, so that, like their craft, they merge with the local scene.

The dhows rarely make a landfall in India. Once a dhow has sailed on its five- or six-day voyage from Dubai, a discreetly worded cable goes to Bombay and a fishing boat sets out from one of the fishing villages like Virar, Dadar or Bandra just north of Bombay. Far out at sea the two craft rendezvous after carefully exchanging coded signals. The golden tola bars have been stuffed into the pockets of smuggling jackets during the voyage from Dubai, and now eighty or more jackets, each containing up to 1,000 tolas, are swiftly transferred. Then, as the fishing boat heads toward shore, the Dubai dhow frequently moves 10 or 20 miles down the coast for a second rendezvous. This time it picks up payment of silver or possibly of currency and traveller's cheques, then turns and sets course to Dubai.

In emergencies, the gold is unhesitatingly tossed over-board, normally attached to marker buoys that will float just below the surface at low tide. The exact position is noted and the dhow can return later to recover the gold or advise the Indian boats of the position.

The fishing boat, meanwhile, will have come in close to shore under cover of darkness and waits until car headlights flashing from the beach give an all clear signal. As soon as the boat grounds, the fishermen leap ashore, laden with the jackets full of gold. These are quickly transferred to the waiting car, which speeds off to Bombay. Once in the city, the driver leaves the car at a prearranged place—often the Regal Cinema or the City Hall—with the key in it. Fifteen or twenty minutes later, another man appears, slips into the driving seat and drives across the city to an empty apartment rented by another member of the syndicate under a false name. There the gold-filled jackets are unloaded and held until individual carriers call to pick up one or two jackets at a time to start the distribution throughout India. At each stage of the operation, the man involved knows only his own precise role. The fisherman does not know the name of the car driver. The first driver does not know the second. Each man along the line receives roughly 1 rupee for every tola he moves. Thus a man wearing a jacket containing 1,000 tolas gets 1,000 rupees ($147). Communications are handled by two-way radio or through public telephone boxes—never from home telephones.

There is, of course, always the temptation for a carrier somewhere along the line to abscond with all the gold. The captain of a Dubai dhow once abandoned his vessel when it rendezvoused with the Indian fishing boat, hitched a ride ashore and vanished with the gold. He had little time to enjoy his loot. The gold network has good intelligence. Soon he was located in a café in Bombay. He was hauled back to Dubai in a sack and, according to some reports, flogged in front of his fellow dhow captains just to discourage them from having similar urges. He then spent many months languishing in the local jail.

The value of the gold going into India each year is so spectacular that bribes can almost always pave the way of the gold route. An Indian customs officer, earning only 200 rupees a month, can earn five months' pay by walking around the corner. If he will not co-operate, drugged coffee or tea can send him into a convenient doze.

'India,' said one senior customs man in Bombay, 'is the most fertile country in the world for the smuggler. You could sell your cufflinks, your pen, your shirt, your tie, even your trousers at a premium because they are foreign made. The whole coast from the Rann of Cutch to Cape Comorin is open. How can we watch it all? We haven't even got fast or seaworthy launches to catch the smugglers' boats.'

Although the customs officials admit frankly that there is considerable corruption within their ranks, the suggestion by some authorities that the customs men receive a specific payoff of 3·7 per cent of the value of the gold smuggled in (assuming they catch one shipment in twenty-seven) simply does not stand up to examination. Individual officers are heavily bribed without doubt, but any formal percentage would be totally impractical, particularly as many senior officers take their job extremely seriously and conscientiously.

'I've been here thirty-two years,' one senior man said, 'and the only reason I'm staying on is to get my pension. If I was getting that sort of bribe I'd have retired years ago.' But they are battling hopeless odds in trying to stem the flow of gold into a nation which has been described as 'a sink for precious metals' for generations.

Over the centuries, gold has become so closely interwoven with the religious and social life of India that no amount of legislation forbidding its import or controlling its holding can quench the demand. India does have two small gold mines of her own. The Kolar Mine in Mysore produces about 2 tons a year, but this is all added to her official monetary reserve. The Hutti Mine turns out a further $1\frac{1}{2}$ tons a year, which is allocated by the Gold Control Administration to authorised users for use in gold thread for saris, fountain-pen nibs, liquid gold and electronics. But this is only a trifling part of the Indian demand for gold, which was at least 200 tons a year by 1970.

Attempts to curtail it have been singularly unsuccessful. In 1963 the Finance Minister Moraji Desai tried to cure the addiction overnight by introducing a bill forbidding the manufacture of the ornaments of over 14 carat (whereas most of the gold in India is traditionally 22 carat). 'I have no doubt that most families in India will welcome these decisions with a sense of relief,' he said. 'All over the world ornaments are made of gold of 14 carats or less. Nine out of every ten families which buy ornaments in our country do so under social pressure and at great cost and inconvenience to themselves. I, for one, do not regard even the demand for gold ornaments as socially justifiable. I regard it as the moral duty of every patriotic citizen to refrain from buying gold in any form.' Not many Indians shared his sentiments. One hundred of India's 450,000 goldsmiths committed suicide, there were protest demonstrations in New Delhi, and a flourishing black market in 22-carat gold soon developed. In the autumn of 1966 the government had to admit defeat. Once again 22-carat gold ornaments could be made, although officially no one could hold more than 171·5 tolas of gold without declaring it.

Announcing this capitulation to gold, Mrs. Indira Gandhi said, 'A measure of socio-economic reform which is aimed at changing the centuries-old traditions and customs cannot be expected to become fully effective within a few years.' The gold habit is, in fact, so deeply inbred in India that only a social revolution will ever change it.

It stems from a simple rural economy in which there is distrust of currency, almost total ignorance of banking and fierce family pride in the value of gold ornaments that can be given to a newly born child or a bride.

'The outlook of the peasant is very narrow,' says J. P. Tiwari, a former President of the Bombay Bullion Association. 'He is always faced with the fear of famine, his crops depend on the whim of the monsoon, he knows nothing of banking or credit. Now what is the one thing to tide him and his village over an emergency? When famine comes they must have something tangible to convert— gold.' Gold rings and necklaces are always lavished on a newly born child and these will be restyled and enlarged as the child grows. For a girl, gold assumes increasing importance. Under Hindu law it was long impossible for a woman to have any proprietary rights over her father's or her husband's property, so some extra provision had to be made for her security in the event of their death. She could always keep her personal ornaments and jewellery—a tradition known as the Stridhana. Thus a bride was always heaped with gold ornaments, and the Stridhana became a walking insurance policy against evil days. The value of the gold bestowed on a bride by her husband and his family was vital to family prestige. 'Their status is judged by the gold they bring,' says Tiwari. In the country, where a girl frequently marries a man from another village (to prevent inbreeding), the honour of both communities was involved. Even the poorest families would endeavour to muster at least one tola of gold; richer communities such as the

Marawis in Rajasthan would expect a family to give at least 100 tolas. The Bombay Bullion Association has calculated that if each of the 120 million families in India has a marriage once every fifteen years and a minimum of one tola changes hands at the ceremony, the annual demand would be 8 million tolas (almost 100 tons) of gold.

Handsome boxes, lined with blue velvet, containing a selection of gold ornaments are on display in every gold-smith's shop. The bride can take her choice of boxes with 4, 8, 12 or even 16 tolas of gold. The 4-tola box, which is the most popular, holds one necklace, 2 earrings, 1 ring and 1 bracelet, all in 22-carat gold; its cost, early in 1972, was exactly 1,000 rupees ($140).

Many young Indian girls nowadays object to this marriage custom, but they are frequently obliged to accept gold ornaments anyway. 'I had to accept 10 tolas of gold,' said one girl in New Delhi, whose ambition is to open a boutique, 'but that's not much. My mother received 100 tolas when she was married. I had to accept the 10 tolas so that both families could keep face, because all our friends want to know how much gold is exchanged.'

The marriage season begins in November, when the first harvest after the monsoon is in, and continues through until June. The rise in demand for gold in late autumn also increases as the farmer puts most of his profits in gold. Since the banking system, especially in the central and southern states of India, frequently does not cover rural areas adequately, the farmer keeps his savings in gold ornaments instead. This is his insurance policy against a poor harvest or an unfavourable monsoon the next year. In the spring, when he needs seed, he goes back to his local goldsmith, who is really fulfilling the role of bank manager, and sells a little gold.

In the last few years the 'green revolution' in Indian agriculture has at once been reflected in the gold markets.

The farmers, using more fertilisers and an increasing number of tractors, are winning a much better return from their land. Many are now able to bring in three harvests a year, instead of two. With good harvests behind them every year from 1968 until 1971, the farmers have had plenty of surplus cash—much of which has immediately been turned into gold. The poor monsoon in 1972, however, was a temporary setback that reduced gold sharply.

The goldsmith himself has also been doing well. Beside increased sales he reaps additional profits because his customers have their ornaments re-styled constantly. He charges a fine fee for the alterations. The local goldsmith obtains his gold, sometimes in the form of the smuggled 10-tola bars, but more often in 10-gram bars which may be legally traded, from various wholesale markets. Bombay, of course, is the prime fountain for gold because of its close connection with Dubai. But some smuggled gold these days finds its way also direct to Amritsar (it is smuggled over from Lahore in Pakistan), Bangalore and Madras. Usually the 10-tola bars are swiftly transformed into ornaments or 10-gram bars to 'legalise' the gold; possession of a 10-tola bar in India is highly incriminating.

The Bombay gold market itself, presided over by the Bombay Bullion Association, is in the narrow tangle of streets known as the Javeri Bazaar. The teeming crowd of the bazaar is so dense that most motor traffic gives it a wide berth; only sacred cows mingle with the thousands bargaining for bracelets and necklaces in the scores of goldsmiths' shops, each with its own tiny furnace in the back room for melting down or refashioning ornaments. On Sheikh Memon Street a solid four-storey building houses the offices of the Bullion Association. Until Moraji Desai's campaign against gold in 1963, the Association's Gold and Silver Exchange boasted a turnover of up to 40 million ounces of gold a year and was the largest market for the forward buying of gold in the world. Desai's gold regulations closed the Exchange and despite the relaxation of the rules on the buying of 22-carat gold, it has not reopened. Nor is its refinery, which used to melt up to a million ounces of gold a year, now permitted to handle gold. For a while, however, the refinery found a lucrative alternative occupation in melting down the silver which flowed out of India in an unending stream from 1965 until 1971. There was a constant parade of people picking their way through the sacred cows in the street outside to the refinery door. Each new arrival clutched an old shopping bag or paper bag stuffed with a jumble of tarnished silver ornaments. These were quickly weighed and dumped into a crucible which vanished into one of three furnaces flaring beneath the refinery floor. Fifteen minutes later, the crucible was withdrawn and the molten metal poured into a neat bar of silver. Once it was quenched, the customer paid a melting and assaying fee of 9 rupees ($1·20), popped it in his shopping bag and vanished into the mêlée in the street outside. The bar was on its way to Dubai by evening.

The average Indian family's holding of gold is probably very small; but given a nation of over 500 million people, the sum total is vast. The Reserve Bank of India estimated in 1958 that 3,150 tons of gold (then worth $3·7 billion at the international price) was held in private hands in India. Allowing for gold smuggled in since that time, the Indian hoard probably reached well over $7 billion (at the monetary price of $42 an ounce) by the end of 1972. Since many Indians regard the Reserve Bank's original estimate as conservative the total may be even higher. As early as 1919 G. Findlay Shirras in his book *Indian Finance and Banking* calculated that the private stock of gold in India was 2,600 tons, while the Gold Delegation of the League of Nations, a decade later, estimated that India took 15 per cent of all the new gold mined between 1835 and

1929. The magnitude of this hoard was confirmed during the 1930s, when famine and depression forced many Indians to sell their gold. In 1932–33 alone, over 240 tons of gold, worth $280 million, was dishoarded; in all, between 1931 and 1940 India disgorged more than $1·5 billion in gold. This loss has been more than recouped.

Since the gold vanishes without any formal record into the sponge of India, it is hard to assess how the hoard is dispersed among the different provinces and the multi-layered class system. One calculation suggests that 90 per cent of the gold is in the coffers of 3 per cent of the population. There are certainly some handsome private nest-eggs. The Nizam of Hyderabad, hardly skimming the surface of his fortune, sold over $1 million of gold through the Reserve Bank of India during the 1950s. One prince whose flat was raided in Bombay in 1963, after the gold control regulations came into force, had $130,300 in gold tucked away behind the wall panels. The turbaned Marawis of Rajasthan, who have long dominated the Indian commercial scene and have a strong hold on India's army of goldsmiths have been stockpiling gold for generations, so that Indians often joke that Marawi homes are built on gold foundation stones. 'Their wealth is unfathomable,' says one bullion dealer with a sigh of longing. 'If you picked up Rajasthan and shook it, it would rattle like Fort Knox.'

The temples have great treasure chests of gold. When the Indian government first tried to regulate the holding of gold in India in 1963 several temples argued that they could not surrender the gold, as it belonged to God, not the temple. Others made token offerings of gold to the government, including the temple at Tirupati near Madras, which offered to weigh the late Prime Minister Shastri against gold and contribute his weight in gold. Shastri, a tiny man, did not accept the offer.

The golden stockpiles of the princes and temples were chiefly amassed during the British Raj. In those days the gold from London was specially polished and wrapped in scented pink paper before dispatch to the maharajahs. Now most of the hoarding of large quantities of gold is by India's *nouveaux riches*—both industrialists and film stars— who do not wish to declare all their income. 'People who make money quickly don't want to show it by buying houses or cars,' explained a Bombay businessman, 'so they buy gold. Some of the film community here in Bombay keep most of their money in gold and diamonds they buy on the black market.'

The only signs of a crack in the gold habit are among the middle-class Indians living in cities like Bombay and Calcutta. Here Hindu women are beginning to throw aside the traditions. Girls go dancing, which they would never have done twenty years ago and they are beginning to question the value of gold. Yet outside these cities, the lust for gold is as strong as ever. 'Gold is a heritage of the uncertainties of the Indian economy,' says a senior Indian economist, 'and the problem is that even over the last twenty years there has been economic verification of its value. The devaluation of the rupee and the slow growth of industry have not made other forms of investment more profitable.' Those who do campaign against gold, which causes a serious drain on India's precious foreign exchange reserves, make little headway. 'If you preach against gold,' says the economist, 'people say "Oh, that dear fool of a professor".' For everyone who argues against gold, there are a hundred to shout its virtues. One former President of the Bombay Bullion Association even goes so far as to call the gold smugglers patriots who are fighting to bring through a commodity vital to the nation. The financial editor of *The Times of India* once wrote, 'It would require all the crusading zeal of a Gandhi to wean the people from the habit.'

Two projects to draw out the gold in exchange for gold bonds met with dismal failure.

The first in 1963, just after the Chinese hostilities, prized only $1·4 million from the hoarders. A second was launched in 1965 during the war with Pakistan, under the grand title of National Gold Defence Bonds. People subscribing gold were offered 7 per cent interest and promised that their identity would remain a secret. The bonds were exempt from wealth and capital gains tax and will mature in 1980, when gold will be repaid. But this appeal to Indian patriotism to contribute to the war effort culled less than $15·5 million in gold—about one-thirteenth of one year's smuggling into the country. In fact, according to the basic psychology of gold hoarding, these were both precisely the wrong moments to make such an appeal. The threat of war makes people clamour for gold, not surrender it for pieces of paper.

One solution to smuggling, if not to gold hoarding, is that the Reserve Bank of India buy gold on the free market at the international price and sell it to the hoarders at a comfortable profit. This would at least put some rein on the drain of foreign exchange and is a policy that was adopted by the Japanese central bank to defeat the gold smugglers. Japan, however, needs to buy much less gold and has a healthier balance of payments than India. An Indian banker points out, quite justifiably, that it would be too embarrassing for India to be receiving vast amounts of aid from the United States, Britain and the World Bank, and then to spend perhaps $150 million buying gold in the London market. 'Some American congressman,' the banker remarked, 'would soon be getting up to know what the United States was helping us for, when we could go out and waste all those dollars buying gold.'

So the gold smugglers flourish, the gold floods in, the precious foreign exchange flows out. From time to time the customs make a spectacular haul—their best was 28,000 ounces, worth almost $2 million at Indian prices, found in two flats in Bombay in September 1966—but as fast as they close one loophole, another opens up.

At one time Air India planes coming in from Beirut had gold concealed in the ladies' toilet and in the engine cowlings. Mechanics at Bombay's Santa Cruz Airport were paid 3 rupees for every tola of gold they removed while they were servicing the aircraft. In 1959 a BOAC stewardess was caught by the customs at Dum Dum Airport, Calcutta, with one gold kilo bar in her bag when she landed from Singapore. While she was being questioned, a BOAC steward who had just come in from Beirut passed the interrogation room and is said to have called out to her, 'Don't trust those Indians.' The piqued customs officers promptly took him off to be searched and found he had 175 10-tola bars in a waistcoat beneath his shirt. In the ensuing investigations more than twenty BOAC stewards and stewardesses were interviewed, although no more were caught carrying gold.

Nowadays Bombay is the focal point for gold smuggling, but until the mid-1960s Calcutta was also an important entry point.

Ships coming up the twisting 80 miles of the Hooghly River from the sea offload gold to waiting river craft, which speed the gold to the Bowbazar Street gold market. The Calcutta gold smuggling is controlled by some of the 50,000 Chinese living in the city, who maintain close contact with relatives in Hong Kong, where most of the gold is slipped aboard ships. Several shipping companies whose vessels ply the route regularly search their ships after they leave Hong Kong or Singapore to try to find gold, rather than face the losses incurred if the Indian customs find gold and impound the ship pending completion of inquiries. Such searches are not always successful.

At first glance the M.V. *Ruth Everett,* which came slowly up the Hooghly in the summer of 1960, was just another small freighter with black hull, white superstructure and chocolate-coloured smokestack. But tucked into her air ducts were over 10,000 ounces of gold worth almost $700,000 on the black market. The first clue to the real value of her cargo came from a Filipino crewman, who was caught by a dock policeman with 156 ounces of gold hidden beneath his shirt as he came ashore. Then the hunt was on. The customs virtually took the ship apart. They found gold in cavities behind bulkheads, gold in the air ducts and even 300 10-tola bars in the ceiling of the sickbay. The ship had picked up the gold in Hong Kong and in their haul the customs found several books of counterfoils from a Hong Kong lottery. The other halves of the tickets had been airmailed from Hong Kong ahead of the ship, so that crewmen delivering gold ashore matched counterfoils with their contacts to ensure they delivered the gold to the right person.

Such a catch boosts the morale of the customs, but makes little impact on the far-flung web the gold syndicates have woven east of Suez.

'Everyone is in smuggling,' says a Bombay customs man. 'There was a time when we used to know it's A, B, C or D who is smuggling, now it's only A, B, C and D who are not smuggling.'

The main customs house in Bombay hardly has the air of a command post waging war against the modern high-speed gold smuggler. There appear to be few, if any, typewriters and no filing cabinets. All files, tied together with tattered ribbon, are stacked loose in high piles in the corners of the lofty rooms of the customs house. Anyone searching for one has to ferret through a 10-foot-high stack. Huge fans turn slowly on the ceiling as the clerks bend over their ancient wooden desks, scratching away with fountain pens. It looks exactly as one pictures the offices of the East India Company in the eighteenth century, and Clive of India seems due at any moment. It is small wonder that even the most hard-working and conscientious customs men seem demoralised.

'The smugglers are smarter and cleverer than we are,' an officer in white ducks says sadly, sipping his tea. The customs' main tactic is to play off the rivalries and jealousies of the different smuggling outfits. 'It's a dirty business of bargaining with people in hotel rooms.' Their information often comes from deserted wives or angry girl friends. It may be an anonymous air letter from a man in Hong Kong whose wife has left him for a sailor. When she flies to India to meet the sailor's ship coming into Bombay, a letter reaches the Customs Intelligence Division suggesting they search the sailor for gold when his ship comes in. The intelligence officer phones the harbour customs office. 'What time is this ship coming in? Six tomorrow morning? All right, send the boat out to meet her outside the harbour. Make a search and look at this chap in particular, but don't let them know we have a tip-off.'

On another occasion the customs men waited outside a hotel room in Bombay for several nights, hoping to catch a young gold smuggler with tola bars on him. There was no sign of gold, but he did have a visit from a girl each night Finally, they decided to try bluff. They raided the room. When the young man denied all knowledge of gold, the officer said, 'We have tape recordings of your entire conversation with this girl over the last few nights. Would you like them played over to her family and your family? Go away and think about it.' In fact they had no tapes at all, but three days later the man came to the customs, announced that he and the girl intended to marry and live in the Middle East. The customs told him he was free to go, and wished him luck. Three months later they started getting information from the grateful young man about other gold-smuggling syndicates.

India's classic smuggling saga, which has taken some new twist each year since 1962, centres on a swashbuckling American airman named Daniel Hailey Walcott, who smuggled everything from gold and diamonds to watches and shotgun shells. Walcott, a broadshouldered Texan, entered the scene as President of Trans-Atlantic Airlines, which operated a fleet of four planes flying freight from Europe to the Middle East, India and Japan. The airline had a contract with Air India, and Walcott himself popped in and out of India all the time on business. In September 1962, a casual search of his twin-engined Piper Apache plane at Safdarjang airport, New Delhi, revealed he was carrying 10,000 rounds of 12-bore shotgun ammunition and 760 rounds of 16-millimetre rifle cartridges, which he had neglected to declare to the customs. Since he had filed a flight plan to Jaipur, the great Indian hunting centre, for the following morning, he was clearly intending to sell the ammunition illegally in that city. He was arrested in his luxury suite at the Ashoka Hotel in New Delhi and the following summer was sentenced to six months in jail and a fine of $420. On appeal, he was let off with the time he had already served in jail. However, during his enforced stay in India he had run up a vast number of unpaid bills. So his aircraft was impounded. That did not hold him for long. Each day he went out to the airport, where he was allowed to run up the engine of the plane to keep it in working order. Every time he put four gallons of petrol in the tank. He never used all four gallons. One day, instead of running the engine for a few minutes, he opened the throttle and took off, despite five airport guards who clung desperately to the wings until the wash of the propellor swept them aside. He headed for Karachi, hotly pursued by Indian Air Force jets, but was safely over the border of Pakistan before they found him. Explaining his escapade to the press, Walcott said cheerfully, 'What I did was to drive down to Safdarjang, walk across to my Piper Apache, jump in, open the throttles and fly over to Karachi. The only violation of Indian law I have committed was to waive procedural red tape because I had more than I could stomach.'

He was soon back in India. In June 1964 he hired a twin-engined Riley 65 aircraft in London, because his own plane had been impounded by the Lebanese authorities, who accused him of making aerial photographs of their military installations for Israel. Walcott was travelling on a British passport under the name of Peter Philby and had recruited another Briton as co-pilot. Flying out to Nicosia they picked up a consignment of 675 Roamer watches and a few hours later zoomed down to land on a wide sandy beach at Murud, 150 miles from Bombay. The plane landed heavily, breaking a nose wheel and buckling the propellers. Walcott and co-pilot climbed out unperturbed. When the local villagers came running up, Walcott told the startled headman to set a guard on the machine while he and his friend went in search of a mechanic. They set off to catch the bus to Bombay, lugging two suitcases containing the watches and possibly some gold. Arriving in Bombay at dawn, they dropped off their bags with an ally and made for Santa Cruz Airport, bluffed their way into the transit lounge and mingled with passengers coming off an East African Airways flight from Dar es Salaam. They showed their passports and health certificates and thus entered India 'legally.' Then they booked aboard a plane for Karachi and were safely out of India before the headman on the beach even began to ponder when they were coming back with a mechanic.

On his next trip, Walcott was more successful. According to Indian police, at least once in August 1965, possibly more often, he dropped gold from the air onto a road near Parsoli 170 miles north of Bombay, where a French colleague, Jean Claude Donze, was waiting to pick it up. Donze, who also used the alias Pierre Carraud, had already tangled with the Indian authorities in 1962, when he was mentioned in the case against an Englishman, Ronald Stanton, who was caught smuggling $45,000 worth of gold into India.

Encouraged by the success of the air drops, Walcott and Donze both ventured back to Bombay in January 1966. They came by way of Colombo and Madras, Walcott travelling on a British passport in the name of Barry Comyn (Barry Comyn, Scotland Yard later found out, was killed in the blitz in London in 1940 when he was six years old). 'Comyn' and Donze checked into Room 503 at the West End Hotel on Marine Lines in Bombay. A couple of nights later, an alert detective, making a routine check of the hotel register and all overseas phone calls made by guests (smugglers often phone contacts abroad to arrange shipments), paused when he saw a number 'Mr. Comyn' had called in Colombo. It had been mentioned in a recent Interpol circular. Reinforcements were rushed to the hotel, Room 503 was raided. Mr. Comyn had $32,500 worth of diamonds taped to the sole of his foot with a plaster and hidden in a sock. There was also a canvas gold-smuggling jacket in the room. Only later did the police realise the true prize they had caught, when 'Comyn' confessed that he was really Walcott. Donze was an even greater prize. Although he is less flamboyant than Walcott, Indian customs men regard him as a key man in gold and diamond smuggling.

The Walcott-Donze saga did not end there. One hot afternoon in the summer of 1966, while they were awaiting trial in Madras on charges of entering India under false passports, the pair made a break for freedom over the wall of Madras Central Jail. Walcott got as far as the top of the wall, but was hauled back by warders. He and Donze were given six months for this attempted escape. This was followed by five-year sentences on the false passport charges. Sentencing them, the judge said, 'These adventures of foreigners are bound to spread like infectious diseases and prove to be a great source of nuisance and danger in future unless they are promptly scotched and nipped in the bud.'

But the misadventures of Daniel Walcott and Jean Claude Donze did nothing to stem the flow of gold into India. Indeed, as the 'great revolution' in agriculture spreads and general prosperity increases slowly, the demand for it will almost inevitably rise.

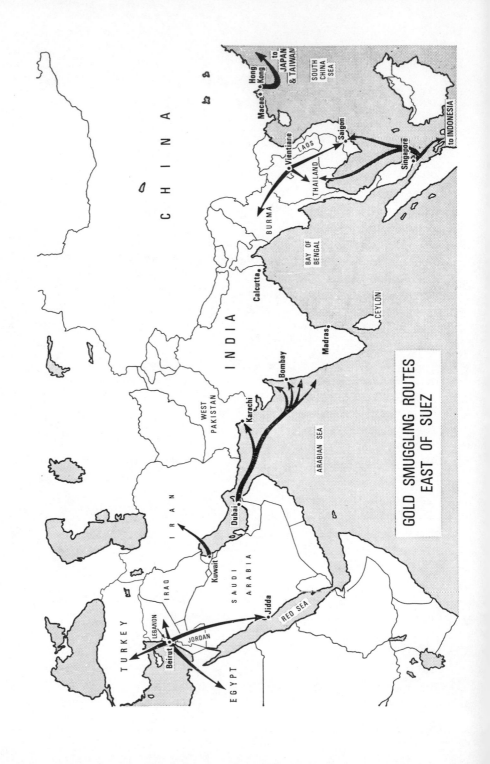

GOLD SMUGGLING ROUTES EAST OF SUEZ

13

The World of the Gold Smuggler

The advertisement in the Personal Column of *The Times* in the autumn of 1967 sounded rather tempting: 'Young men of initiative wanted for job with rewarding travel opportunities.' It sounded like a nice chance to get away from the fast approaching English winter. A host of people replied—including a handful of sales assistants in Harrods anxious to escape the mounting seige of Christmas shoppers. Their enquiries lead them to a somewhat seedy flat in South Kensington, where an urbane Old Etonian explained what the caper was all about. How would they like to take 40-kilos of gold (then worth about $50,000) to Hong Kong for him? Naturally all expenses would be paid and at the end of the jaunt £200 would be paid into their bank account.

The Old Etonian had not simply been reading Ian Fleming's *Goldfinger*; he was embarking on a serious attempt to win a slice of the international gold smuggling traffic. While the notion of gold smuggling today may sound a trifle romantic, in practice more than $750 million worth of gold—almost half the world's annual production—passes along smuggling pipelines somewhere on its journey to the final customer. The Dubai boats stealing over the waters of the Arabian Sea to India with upwards of $300 million a year (see preceding chapter), are but one link in a golden chain. For every ounce of gold bound for India, another is being spirited into Brazil or Japan, to South

Vietnam or Turkey, to Morocco or Indonesia. Wherever the import of gold is forbidden or strictly controlled the smuggler is not far away.

Whatever the ultimate destination, the ingenuity in hoodwinking authority is unbounded. Gold pellets are substituted for date stones by Moslem pilgrims returning home to Pakistan from Mecca. Gold travels hidden in bicycle frames; salmon and chicken bear golden eggs; the endless folds of Indian saris or the saffron robes of a Buddhist priest offer ideal concealment. Tins of milk or even motor oil are excellent, because the gold does not rattle. Journalists have even carried typewriters with solid gold space bars. One man, smuggling gold from the Philippines to Hong Kong, had tiny chips of gold hidden under 2,642 strips of adhesive plaster stuck all over his body. Small gold bars can easily be carried internally; women, they hasten to point out in smuggling circles, can carry twice as much as men. 'You can't imagine the fertility of these people's minds,' says an admiring Swiss bullion dealer.

The rewards are enough to exercise anyone's imagination. Most smugglers operate on a rule of thumb that gold smuggling is profitable if the margin between the price of gold, where they purchase it and its eventual destination is $75 to $100 per kilo. The potential gross profit, therefore, is a staggering $45 million *each year*.

With so much at stake, gold smuggling is not a business to be dabbled in. In fact, over the years a variety of extremely efficient international syndicates have grown up, whose turnover, if declared, might well place them in the world's top 500 companies. These syndicates, often employing a hundred or more couriers, have worked out of Beirut, Geneva, Brussels, Vancouver, and, as the advertisements in *The Times* revealed, even London.

The structure of a syndicate is often as clear-cut as any

well-regulated company. At the top is the leader, the managing director. He works discreetly behind the scenes all the time, planning, making high-level contacts both on the gold-buying and gold-distribution side. If there are problems in Tokyo or Rio, he is aboard the next plane to try to smooth things over. His team will normally be operating to four or five different countries so that he can play the current market conditions in each. He has to be careful not to send too much gold to any one market and so push the price down. He rarely carries gold himself and will always behave with the utmost propriety. In India, for example, he will make quite sure he cashes all his traveller's cheques at the bank and does not have any dealings with the currency black market. A senior man who once broke this rule, some years ago, proved its value. He was carrying 75 kilos of gold worth almost $85,000 from Beirut to Hong Kong. His plane was held up by engine trouble in New Delhi, and the airline, to keep the passengers amused, announced it was taking everyone on a tour of the city. The smuggler said, thank you very much, but he would prefer to stay at the airport and rest. The airline was persistent; the plane would be a long time and he must come into town. Finally, he had to agree. Naturally, said the airline, all the passengers would have to clear customs. He was caught with the gold beneath his shirt. Today he has retired from the business and runs a hotel in Lebanon. Since then all the leaders adhere to the golden rule of never carrying gold themselves.

Each leader normally has several lieutenants. One is in charge of accounts. His responsibility is to make sure that gold payments come through promptly. He must be an expert on currency black markets and be able to devise a new method of getting money out of a country if the authorities stop up one loophole. Working closely with him is the 'vice-president' in charge of travel arrangements.

He is a walking airline timetable and can conjure up three different ways to get an urgent order of 40 kilos to Tokyo. He must be on intimate terms with all the local airline managers and may play off the different lines to give him a discount. He often operates through the front of a travel agency. Another executive is the personnel manager, responsible for the recruitment, training and day to day work of the gold carriers. 'He has to listen to keep the carriers happy,' a smuggler explained. 'He has to listen to their complaints and be very careful that a carrier doesn't suddenly become too rich.'

Each syndicate also has 'correspondents' or 'station managers' in strategetic cities to handle local arrangements. Besides acting as liaison man with the syndicate's clients there, the station manager has to ensure a smooth arrival for the couriers. He must know the layout of the airport intimately, because couriers may be able to leave their gold in a toilet in the transit lounge, where it is picked up later by a cleaner. Alternatively, in Hong Kong he will know that passengers who have a few hours' wait can leave the airport for a quick look round, without their baggage clearing customs. So provided they are carrying the gold in a special smuggling jacket beneath their shirt, they can often walk through arrival controls with only a cursory check. The station manager will also consider the possibility of bribing local customs men—a task that requires considerable thought; it is not much use bribing one man as the courier may not end up in his line. Ideally a whole shift has to be corrupted to ensure a smooth passage. Then each courier can be decked out with some special identification mark—usually a tie—which the customs officers will spot and obligingly wave him through. A tie also helps the station manager pick out incoming couriers in a crowd. One syndicate used a wine coloured tie, with black diagonal lines forming a diamond pattern, while within the black diamond silver lines formed a smaller diamond. A natty job that any old boy network could be proud of.

The actual extent of 'buying' customs men is hard to determine. Gold smugglers talk a lot about it, but that may often be bravado—or just a nice way of making a little extra on the side themselves. A station manager may claim to be paying the customs at Djakarta airport $2,000 a month, when he is either paying them only $1,000 or nothing at all. He can hardly ask them for a receipt to submit with his expense account.

The ideal, of course, is to be able to bypass customs altogether. A diplomatic passport naturally comes in very handy, as baggage is usually exempt from customs' searches. Several Middle East, African and Asian diplomats have developed profitable sidelines globe-trotting with their briefcases bulging with gold instead of state papers. 'I know a Philippine diplomat,' recalled a Hong Kong exchange dealer, 'who used to come in and out of here thirty times a year with gold.'

State visits are even better cover. The syndicates working out of Beirut have often cashed in on a tour of India, Pakistan or the Far East by some Middle East ruler, since both he and his entire entourage pass unimpeded. The customs may know full well that the bags of their disting-uished visitor's aides are stuffed with gold, but they cannot act. A major diplomatic incident over gold smuggling as the festitivities commenced would hardly be conducive to strengthening the cordial ties a state visit is designed to create. So Beirut is full of tales of Saudi Arabian princes taking $1 million in gold to Japan, or a shiekh from the Persian Gulf arriving in India with $500,000 in gold bars in his suitcases. It is even rumoured that one Middle Eastern nation dispatched almost solid gold furniture to equip its embassy in New Delhi.

Such smuggling by 'royal warrant', however, is infrequent

The regular donkey work in each syndicate is done by the couriers. There may be as many as a hundred of them with some loose working arrangement to each group. Originally the majority were Syrians, Lebanese, Jordanians and Iraquis recruited in Beirut, when that city was the prime launching pad for gold smuggling throughout the Middle and Far East. They still operate the short runs around Beirut to Turkey, Syria, Egypt and Iran, but farther afield customs men now regard them with considerable suspicion. The recruitment net has thus widened to include West Germans, Britons, Swiss, Italians and Swedes. Businessmen and airline crews whose normal work necessitates their travelling extensively, make ideal couriers as they have a legitimate reason for visiting odd corners of the world many times a year. Students, normally poverty-stricken and tempted by the prospect of free travel and pay, are also frequently enlisted. One group developed a speciality for employing Catholic priests—both genuine and fake—on the sound theory that clergymen are less likely to be searched closely by the authorities. His clothes—or robes —also offered better concealment for a canvas jacket stuffed with gold than an executive's well-cut Savile Row suit.

A Geneva syndicate relied for many years on German taxi drivers from Frankfurt, Cologne and Munich, who seemed to enjoy the occasional few days off from traffic jams for a jaunt to Bangkok or Tokyo for a $500 fee. A courier manager in each city handled the recruiting and acted as the essential link with Geneva. Usually the couriers went over to Geneva to pick up the gold and were given a night out on the town before departing to steady their nerves. Sometimes they just picked up the gold in the transit lounge at Frankfurt Airport. A man from Geneva would bring the gold in and rendezvous with the outgoing courier in the toilets of the transit lounge, and pass the gold over the partition between the two cubicles. Thus neither man ran the gauntlet of German customs with the gold.

Every group interviews its couriers thoroughly before taking them on. They will be asked what previous work they have done and why they left it? Do they have a criminal record?

'It is most important to have faith in a carrier', said a Beirut gold dealer, 'Crime never comes into this business and we don't want people with a criminal record.' The syndicates are constantly at pains to stress that they do not regard themselves as criminals, but that they are merely supplying a genuine demand that is stifled by the economic regulations of some countries. They try to keep clear of any association with the international drug traffic.

If a new courier's credentials are approved, he will go into training. The crucial thing is to become accustomed to wearing for six or seven hours at a stretch one of the canvas jackets in which the smugglers carry the gold beneath their shirts. These jackets have pouches back and front designed to take either kilo bars, the size of a half pound chocolate bar, or 10-tola bars, which are smaller than a match booklet. A strong courier can carry up to 40-kilos of gold, but it requires considerable practice before he can walk properly. Like a medieval knight being strapped into his armour, he has to be helped into the jacket.

'The first time I put on a jacket with just twenty kilos I nearly fell flat,' a tall, broad-shouldered former courier recalled. The courier has to learn to walk very steadily and deliberately—and at all costs try to avoid bumping into anyone, because the gold makes him top-heavy. One group even bought a row of airliner seats so that their men could practise sitting in them for hours on end, bowed down with gold, without getting cramp.

That Old Etonian in London, signing up the Harrods

staff who answered his advertisements, not only put his
men in strict training, but even wrote a 'Smugglers'
companion' of do's and don'ts. It proscribed, 'he must have
belt or braces, new shoelaces, needle and cotton, safety pin
and buttons.' This was first aid equipment in case he popped
a button en route and so might reveal the golden corset
beneath his shirt. The fresh shoe-laces were vital, because
if an old one broke the courier could not bend down ele-
gantly to fasten it. Each man was equipped, too, with
Dextrasol tablets for extra strength. The manual also
ordered, 'Always dress "English" with collar and tie. Avoid
casual gear. At airports nobody ever gets stopped unless
they draw attention to themselves. Be polite and relaxed
with officials, but never humble.'

Since the private holding of gold is forbidden in Britain,
the couriers for the London operator usually picked up their
golden waistcoats in Brussels (the private holding of gold
being permitted in Belgium). They stayed overnight at a
comfortable flat in Brussels while they were briefed and
checked out on everything from shoe-laces to needle and
cotton. But they were not told until the very last moment
precisely where they were going—just in case they
got drunk in a night club and gaily boasted 'I'm the man
with the golden waistcoat bound for Bangkok tomorrow.'

Regular couriers are usually equipped with two or more
passports, for the authorities are naturally likely to be
suspicious of anyone who flits in and out of their country
too regularly. But their chances of getting caught have
never been very high. Although the ones who are seized
often make the headlines, that is but the tip of the iceberg.
Just before the 1968 gold crisis one group was sending 25
couriers every week to South-East Asia, sometimes with
as many as eight on the same flight. That was by no means
a record. A Beirut smuggler likes to recall how some years
ago he once went out to the airport to watch the Panam

flight to Hong Kong take off with 16 couriers aboard with half a ton of undeclared gold beneath their shirts. There was some speculation on whether the aircraft could take off; it did.

A good syndicate will even look after the family of any courier who is caught. In Germany the courier's wife will often receive about $100 a month pension if her husband lands up in some foreign jail. After all it pays an organisation with a turnover of perhaps $50 million a year to keep its employees happy.

The carrier works in relative isolation. He is one link in a complex chain and he hardly knows the links on either side. His movements are directed by plain language codes, which cannot be broken. 'A cable may say, "Regret to inform you Johnny died last night,"' says the retired gold courier in Hong Kong, 'which means go to see Johnny at such and such an address. You can't crack that kind of code unless you have the book.' The simple word 'Tony' in a cable may reveal to the initiate that 40 kilos of gold is coming in on the SAS flight over the Pole to Tokyo. The carrier will normally be told on departure to go to a specific hotel and wait until he is contacted or possibly to phone a certain number on arrival. He will also be given a password, usually just a Christian name.

The case of Christian Suren Aspelin, a gold smuggler arrested by the Japanese police in 1965, reveals the thoroughness of the syndicates' arrangements. Aspelin was a twenty-eight-year-old Swede and according to his passport was a university student. He visited Japan at least five times in May, June and July 1965, never staying more than two or three days. On August 1 he arrived once again at Haneda Airport, Tokyo, on an SAS flight from Beirut. He had 36 kilo bars beneath his shirt. The customs, acting on a tip-off, decided to follow him. As Aspelin hurried out of the airport and jumped into a taxi, there was a plainclothes policeman right behind him. He was trailed to a hotel in Akasaka, where he contacted a Lebanese named Raymond Joseph Naccachian, who had arrived a couple of days earlier, without any gold, but with his Lebanese girl friend. Aspelin went to the bathroom, stripped off his clothes, took the golden jacket off his aching shoulders and handed it over. Naccachian paid him $560 in cash and Aspelin departed to enjoy the Tokyo night life for a couple of days (instead of following his normal plan of taking the first available plane back out).

Naccachian, who had bought a beat-up Japanese car for $150, waited until the following afternoon, when a cable from Beirut delivered to an English liaison man in Tokyo told them to go ahead and pass the gold on down the pipeline. The two men and their girl friends set off in the car for Waseda University. They parked near a telephone booth outside the main entrance and waited in the harsh summer heat with the gold kilo bars stacked neatly beneath their feet. Eventually a hunchbacked Japanese approached, and knocked on the car door. A password was exchanged. Naccachian told the girls to take a short walk. The Japanese climbed in the back seat and the bargaining for the gold began. He had only 10 million yen (about $33,000) with him, so it was finally agreed he should have 20 kilo bars. That left Naccachian with 16 bars still to sell. He headed back for his hotel; the English liaison man went off to cable for further instructions. But the police moved in and arrested them.

The strength of the major syndicates has been their ability to withstand such losses. 'One group here has a working capital of $3 million,' explained a Beirut dealer, 'but they could take a loss of $400,000 without having to touch this capital—it would all come out of profits. We once lost $70,000 in gold and $100,000 in currency in Syria, but it was all paid for from profits.' The profits have also enabled many senior gold syndicate executives to retire comfortably. 'People here have made their millions and now they've gone into hotels, land and factories,' the dealer added.

In recent years, however, the gold smuggling groups have not had things all their own way. The higher, fluctuating gold price since 1968 has made their task much more difficult. Previously gold had cost $35 an ounce for over thirty years. The syndicates knew they could buy at $35 and get $39 in Hong Kong or $50 in India or $57 in Japan. But the fluctuating price made things much more difficult; a slight drop in the international gold price while the courier was en route could wipe out the profit.

A further complication since 1970 has been the increase in airline security because of hi-jackings. On many routes it is simply no longer feasible to send couriers because of personal searches before passengers board aircraft; that golden waistcoat would be detected instantly.

One thriving little gold market, the island of Curacao in the Netherlands Antilles just off Venezuela, has been stopped completely by airline searches. Up to 1969 Curacao provided a highly convenient jumping off point for couriers bound for Brazil and Argentina. But when the spate of hi-jackings to Cuba began, every passenger going through Curacao was checked. After two Argentian couriers bound for Buenos Aires had been thus ferreted out, the gold traffic stopped abruptly. The trend now, especially to South America, is to send more gold by airfreight with false documents.

The Beirut and European syndicates' markets have been considerably undercut also by flourishing new centres of gold smuggling much nearer the eventual destination of gold. Dubai, as we have seen, has cornered most of the business for India with its dhows. Only a little gold now trickles into India from Europe or Beirut, chiefly by way of

Teheran and Afghanistan. Further east Vientiane in Laos has won much of the South East-Asian market, while the new market in Singapore took over almost all the gold smuggling into Indonesia, which had previously been supplied for a decade by Beirut, Hong Kong and European syndicates.

Within the Far East gold smuggling is dominated almost exclusively by Chinese. They control not just the distribution from Hong Kong, but also the local markets everywhere from Djakarta to Bangkok and Saigon to Singapore. There is an endless family octopus with younger brothers in Surabaja and Taipei, cousins in Penang and Kuching, or uncles in Phnom Penh and Vientiane. Once the gold arrives in the Far East the Chinese take over.

The most informal smuggling centre to have developed is at Vientiane, the capital of Laos. The run-down former French Colonial little town is strategically placed in the heart of the South-East Asian peninsula. To the south are Thailand and Cambodia; to the east South Vietnam; to the west Burma; and to the north, the giant of China. The thick jungles and lonely border posts have made it a paradise for smugglers—not only of gold, but of opium which filters down through Laos and northern Thailand from the Shan states of Burma. Some of Vientiane's gold is undoubtedly used to help pay for the drug traffic. But the real reason for Vientiane's success was the war in South Vietnam that flooded the area with dollars—which were quickly turned into gold.

In its heyday between 1965 and 1968 gold, always in convenient kilo bars, flowed into Vientiane by the plane load. The bars wrapped in newspaper were tucked away into every square inch of the daily Royal Air Lao or Thai Airways flights up from Bangkok (where it arrived in transit from London and Zürich). Passengers even found gold stuffed under their seats. Each flight was met by the local manager of the Banque de L'Indochine, who backed his car right up to the plane. The gold was transferred to the trunk of his car even before passengers' baggage was unloaded. Then the bank manager drove to the customs shed, where the entry of gold was formally recorded. The incoming gold was charged an import duty, which provided one of the main sources of revenue for Laos for several years. With the paperwork complete, the gold buyers crowded round the car. The cartons containing the gold were opened up and it was distributed immediately to customers who stuffed it into tattered suitcases, bamboo baskets and straw shopping bags, and sped away in cars and taxis.

Most of it landed up either in Bangkok or Saigon. The route to Bangkok was very easy. The smuggler just took a little ferry boat across the Mekong River, which flows past Vientiane, to the Thai border town of Nong Khai. There he tucked the gold away in a car and drove the four hundred miles down to Bangkok. One man used to make the round trip twice a week, carrying fifty kilos (then worth about $55,000) each time. Since the price difference between Vientiane and Bangkok was usually at least $75, that meant a nice little profit of $7,500 a week.

Saigon was even more of a gold mine. The price margin was often up to $150 a kilo; the gold flowed down in a torrent. Some of it went by road through the little town of Pakse close to the Laos–South Vietnam border very close to the Ho Chi Minh trail. Most of it, however, was whisked down in Royal Laotian airforce or South Vietnamese planes on 'goodwill missions'. They carried up to half a ton of gold on a single flight, with the full connivance of senior military officials. Everyone was in on the game. A Laotian prince, off to Europe on holiday, went via Saigon so that he could deliver 500 kilos of gold on the way.

The gold was paid for in American greenbacks, which poured into Laos as fast as the gold left.

The spate of dollar bills was so great that the Banque de L'Indochine's branch in Vientiane opened up two special counters just to receive them. Since many of the notes were smuggled into Laos wrapped in polythene bags hidden in tins of fish sauce—a delicacy exported from South Vietnam —the smell in the bank was terrible.

Laos' golden days, however, appear to have ended with the running down of American involvement in South Vietnam. With only a handful of Americans left to spread their largesse among the bars and prostitutes of Saigon, the supply of dollars has largely dried up. While the tremendous black market in stolen American equipment, whose profits were salted away in gold, has also declined. Vientiane's gold smuggling seems to be passing into history.

The outlook for the gold smuggler in several other previously lucrative areas hardly looks cheerful. Indonesia, which suddenly blossomed into a gold smugglers' dream in the economic chaos that followed President Sukharno's fall, is overloaded with gold. For a while the Indonesian demand was so strong that it got the new Singapore just across the water off to a flying start. Although Singapore only began an official gold market in 1969, by 1970 it had become the number one market in the Far East, quite eclipsing both Hong Kong and Vientiane. The Singapore banks licensed to deal in gold were filled with people strapping on golden waistcoats and setting off in junks for the myriad ports of the Indonesian archipelago. In 1970 alone they spirited at least 75 tons of gold into the country almost undetected. The next year, however, the boom petered out.

Nor are things so bright for the smugglers in Japan, which was long one of the most profitable markets. For many years the Japanese allowed no official import of gold and jewellers there had to get by with the inadequate local production. Moreover, the government maintained an artificial gold price of $57 an ounce for their jewellers and industrial users of gold. While gold remained at $35 an ounce, that margin of $22 on every ounce was enough to tempt anyone to a little smuggling. Thousands tried. Couriers flooded from Beirut, Geneva, Brussels, Hong Kong, Vancouver and even by the roundabout way of Mexico City. The trans-Polar flights were a favourite for they cut down the time the courier had to be strapped into his golden corset.

The Japanese customs, perhaps the most efficient in Asia, were ready for the challenge. They devised a gadget— a form of metal detector—that could sniff out the gold as the couriers came through the airport. However, the machine was apparently reliable only at very close range and was not infallible. But it detected several gold runners at Tokyo Airport, so everyone switched to Osaka airport, where it was not installed.

The potential size of the Japanese market, which, by the mid-1960s required at least thirty tons of gold a year on top of local production, made the smugglers more ambitious. They began to devise ways of sneaking it in by the ton. The ultimate caper, in 1967, was to hide the gold in the false bottoms of drums of motor oil being shipped from Canada to Japan. Initially the gold was consigned to Geneva to Vancouver, where, because Canada allows private gold buying, it was picked up quite openly from a bank. It was then slotted into the specially constructed oil drums and loaded into freighters in Vancouver and Seattle. The trick worked for months, until the Japanese customs nabbed a consignment of 100 drums, each with a kilo bar stowed away. Thereafter they vetted every oil can coming into the country; within a week they had prised another 460 kilo bars from oil drums that had been on the high seas before the smugglers learnt that their hiding place was known. The seizure capped a vintage year for the Japanese customs; in all they caught over 3,000 kilos worth over $3·5 million.

But it was the Bank of Japan, rather than the customs'

vigilance which really curtailed the illicit gold traffic. The Bank decided, with excellent logic, that if the smugglers could profit from the difference in the gold price between Europe and Japan, so could they. So in 1967 they began buying gold at $35 an ounce on the London gold market, airfreighting it to Tokyo and re-selling it at the Japanese official price of $57 an ounce to authorised gold users. The government pocketed the difference. With a ready supply of gold available from official sources, many of Japan's precious metal firms began buying a large part of their requirements from the Bank of Japan instead of on the black market. So that although the profit still existed for smugglers, the size of their market fell. Whereas up to 1969 the smugglers handled perhaps two-thirds of Japan's gold requirements, by 1972 their share had tumbled to less than one-fifth. After 1 April 1973 the smugglers seemed to be cut out altogether, for from that date the free import of gold into Japan was permitted. Anyone in Japan can now obtain a licence from the Ministry of Finance to buy gold directly on the London or Zürich markets. So Japan, once the gold smuggler's paradise, has become a legitimate buyer of gold. All those poor 'students' will have to find some other reason for a free trip there.

Japan, of course, is in a unique position. She has a healthy economy and enormous reserves of foreign exchange. She can easily afford to buy a couple of hundred tons of gold a year for her industrial users—or even for gold hoarders. But for India, or Brazil, Egypt, Turkey, Morocco or even Iran, where foreign exchange has to be conserved to buy essential raw materials or industrial goods, then controls on gold imports are inevitable. While that situation prevails the prospects for the gold smuggler, whatever his temporary setbacks, look good.

14

The Future of Gold

The American humourist Art Buchwald, reviewing the steady drain of gold from Fort Knox a few years ago, came up with his own inimitable answer for the future of gold. The United States, he suggested, should end its allegiance to the metal and declare itself to be on the junked car standard. This would solve not only America's monetary problems, but would help to beautify America by collecting in all the old cars desecrating the landscape. The new standard, according to Buchwald, would be simple to implement. Americans would be given thirty days to turn in their junked cars to the nearest Federal Reserve Bank. There the cars would be crushed to the size of a shoebox by one of those giant presses made famous in the film *Goldfinger* and now widely used in scrap yards. The United States army, meanwhile, would be dumping all the gold from Fort Knox into a nearby river to make room for the bar-sized junk cars. Since America junks cars much faster than any other nation, their reserves would inevitably be the largest; the Russians and the Chinese, who have hardly any cars, would be nowhere in the running at all.

Well, at least all bankers, economists and politicians might agree unanimously to junk Buchwald's proposition, which is more than can be said of their efforts to agree on the future role of gold. While they would probably concede that gold is at a vital crossroads in its six-thousand-year history, they hesitate on which way to go now. Straight on along the present path towards the eventual demonetisation of gold? A detour to right or left with gold linked to reserve units like Special Drawing Rights? Or turn round and head back to the so-called good old days of the pre-World War I gold standard? Just now the lights at the junction are on red while the International Monetary Fund's 'Committe of Twenty', created in the autumn of 1972, thrashes the whole thing over. Plenty of other people are chipping in with solicited and unsolicited advice. 'Double the price'; 'the IMF should sell gold to push down the free market price'; 'we need a floating monetary gold price'; 'switch to the SDR standard'. According to your fancy you can line up with Jacques Rueff, Sir Royal Harrod or Dr. W. J. Busschau (to name but a few) going through the gold standard lobby or join Professor Robert Triffin, Edward Bernstein, Fritz Machlup and Fred Hirsch trekking down the demonetisation trail. The latter group are clearly favourites with the Americans, the British and many pundits, but the votes are not all in yet.

In the nature of things there is hardly likely to be a sudden final curtain for gold as a monetary metal (it took ninety-nine years for the demonetisation of silver to be fully implemented), but the trend over the last half century or more has been to push it gradually to one side in favour of currencies and SDRs. Gold no longer makes up 90 per cent of monetary reserves as it did in the 1930s, and is little used to settle international debts.

Paul Jeanty of Samuel Montagu, the London bullion dealers, pointed out at *The Financial Times* conference on gold in 1972 that the history of gold over the last generation has gone in seven-year cycles; from 1947–54 there was a shortage on the free market and prices fluctuated widely; from 1954, when the London gold market reopened, until 1961 the price was very stable and demand fairly even; in 1961 the international gold pool was created and it lasted precisely seven years until March 1968. If the pattern persists look for the next milestone in 1975; what will happen to the two-tier market then?

But before considering the likely long-term role for gold, it is important to know what is the prospect over the next generation for the supply of gold itself. After all, if the mines are running out that will have tremendous bearing on the metal's future.

Gold mining itself is also at a crossroads. For the first time in almost forty years the price has gone up. Will this have the same effect of stimulating production that the last big jump in 1934 had? Remember that both in Canada and the United States record gold production was achieved in the late 1930s because of the encouragement of the 1934 rise from \$20·67 to \$35 an ounce. The trouble this time is that the price increase is too little too late for many mines, both in South Africa and North America. They have hung on so long that the costs of reopening old workings are often prohibitive. Moreover, although the free market price in early 1973 was over \$90 an ounce, there was no certainty that it would stay that high and, even if it did, it was a tiny rise compared with most other metals. As Robin Plumbridge, executive director of Gold Fields of South Africa, pointed out to the American Mining Congress in San Francisco in 1972, the increase in the price of gold between 1934 and 1967 was nil; while silver rose 223 per cent, copper 354 per cent, lead 263 per cent, zinc 233 per cent and tin 194 per cent. Faced with the fixed price of gold for so long, most countries kept their mines going only through substantial subsidies; Canada, the Philippines, Rhodesia, Japan and Australia were all underwriting their gold mines during the 1960s and even South Africa started helping out its marginal mines in 1968. The Soviet Union, too, has clearly been producing gold at an uneconomic rate for years. The lone country, in the major world league, therefore, which was still producing gold exclusively at \$35 an ounce up to 1968 was the United States. Everywhere else subsidies were giving the mines as much as $50 an ounce. The subsidies, however, were usually geared simply to keep existing mines—and jobs—going; they were not large enough to encourage exploration for new ones. The industry has consequently been marking time for many years.

The price rise since 1968 has undoubtedly lengthened the lives of some existing mines, which can go back and dig out the low grade ore they were forced to bypass as long as gold remained at $35. In South Africa, for instance, up to 1971 only ore that yielded over about 8 grams of gold per ton was really profitable, but the higher prices of 1972 made ore with 5·4 grams a ton payable. The lower grade means, of course, that total production is actually smaller on an annual basis, but that the mines can survive much longer and their total yield over the years will therefore be greater. The overall effect on South African production is that it has declined slightly, but remains on a fairly stable plateau, which should extend for more than a decade. This contrasts sharply with the outlook as recently as 1966, when the Chamber of Mines undertook a survey into the long-term future of South African gold mining, which revealed a precipitous cliff of falling output within a matter of years. Nowadays the Chamber of Mines talks of a gentle decline. Its precise steepness will still depend on the gold prices, for costs are rising by at least 4 per cent a year, but at least for the first time in many years there is a chance that the price will keep pace.

An additional bonus may come along if the price of uranium rises during the 1970s. Ten mines contribute uranium as a by-product of their gold operations; while one mine, West Rand Consolidated, is rated as a primary producer of uranium with gold as a by-product. Uranium prices have been in the doldrums since the 1950s, when the United States and Britain, anxious to build up strategic reserves, bought large quantities from South Africa. Uranium production there reached a peak of 5,900 tons in 1959. But once strategic stockpiles were high, the demand trailed off pending a major switchover to civil nuclear power stations. This change has been much longer coming than the South Africans predicted and uranium production slackened to 3,800 tons in 1971. However if, as anticipated, uranium requirements increase in the late 1970s this could be a useful extra stimulant to gold production.

In the meantime, it is the higher gold price that has really caused smiles in Johannesburg. In the autumn of 1972 the Chamber of Mines was cautiously unveiling its thoughts on the prospects for gold production well into the 1980s. They differed radically from its 1966 predictions, which had forecast that output would plummet from the present 1,000 tons level down to 400 tons annually by 1980 and under 200 tons by 1984. The Chamber's president R. C. J. Goode outlined three potential variations on output. The first envisaged no official gold price increase, but a steady rise in the free market price from the present average of $55 an ounce (which is what the mines actually got from limited South African sales) to $120 by 1987; this could hold production at over 800 tons to 1980, with a tail off to about 650 tons by 1987. The two alternatives considered were the monetary price raised at once either to $70 or $100, both of which would also hold production at over 800 tons until well into the 1980s.

However, with the prospect of only one or two new mines being developed in the meantime, output would still fall off quite sharply from the late 1980s onwards. The general consensus seems to be that by the end of the century South African production will be substantially lower and confined primarily to a handful of high grade 'super mines'. As Michael O'Dowd, of Anglo American summed it up; 'Essentially we have thirty years as a major industry, then perhaps another thirty years with only three or four mines operating. Even uranium can only lengthen the plateau before the decline; the drop must come.'

The important thing is that no one is talking of an *increase* in South African production. As her output gradually diminishes, can anyone else make up the leeway? The most obvious candidates are the Russians. Their production, as we have seen, is still a closely guarded secret, but all the indications are that it is over 200 tons a year and rising. Furthermore, a higher price for gold could stimulate the expansion of the industry in the Soviet Union. It is on the cards, therefore, that over the next decade increased Russian production could make up for the slight South African decline; between them they might muster a steady 1,100– 1,200 tons a year.

Beside these two giants, other gold producing countries are so insignificant that even if their output doubled it would have little impact on the world scene. In fact, in most of these other nations the price rise will probably simply slow their past decline. The exceptions are countries like Australia where the gold by-product of the vast Bougainville copper mine will step up total yield sharply. The price rise may also trigger off some increase in other countries where output has really been ignored for years. 'The trouble is,' said a leading geologist at one mining finance house, 'that there has been no money or incentive for exploration for years. With the exception of the United States, most countries are still relying on surveys done in the 1930s when the gold price last went up. In Europe, for example, there has been no attempt to search for gold since then.' This is not to suggest that some major gold field has been overlooked, but it is worth remembering that in the United States the Carlin and Cortez gold mines in Nevada were opened up in the 1960s only because of advances in technology which made it possible to detect and extract microscopic particles of gold. Both those deposits of gold were missed by early gold rushes.

So it is quite likely that if the gold price holds up, there may be some expansion in gold production in countries at present very low in the world league table. Furthermore, other vast base metal projects, like the Bougainville copper mine, will undoubtedly pay an extra dividend in gold. After all, 40 per cent of American gold now comes from base metal ores. While this will almost certainly not be enough to compensate in the long term for a run-down of the South African industry, it should help maintain total world output at around 1,400 to 1,500 tons until well into the 1980s.

There is one dark horse—gold from the sea. Every ton of seawater contains approximately 13 micrograms of gold. Professor Ernest Bayer of Tubingen University has even successfully recovered 1·4 micrograms of gold from 100 litres of sea water in the Bay of Naples, by using an amino-phenol type of chelate polymer containing sulphur. At this extraction rate almost 19 million tons of gold could be won from the oceans of the world. So far, however, no one has come up with a way of extracting it economically on a commercial scale. Even if they did, it would produce gold in such profusion that it would ruin its use as a standard of value.

Realistic planning, therefore, has to make do with the modest 1,500 tons or so annually from conventional sources. But that, of course, is hardly enough to meet the demand for fabrication, which in recent years has been running neck and neck with production. While a higher gold price will slow the fabrication demand temporarily, it should not detract from the long-term upward trend in consumption.

There has been considerable argument as to how accurate the estimates of Charter Consolidated and Consolidated Gold Fields into the non-monetary uses for gold really are.

Clearly in some countries, notably what bullion dealers love to call 'traditional Eastern hoarding markets', it is difficult to be precise and the demand there may fluctuate widely from year to year. But the real key is in the United States, Western Europe and Japan; these developed countries take some 800 tons of gold a year for jewellery, dentistry, electronics and other industrial uses. And their consumption is rising (very fast in Japan) and is relatively unaffected by the gold price.

Gold is not usually bought with any hoarding motive in these countries, but satisfies the taste of an affluent society for handsome jewellery, gold cigarette lighters and even such mundane things as class rings or is used in sophisticated technology. It is reasonable to suppose that in the less developed nations gold consumption will also rise as their economies strengthen. But their motive for buying gold, which is now primarily a fairly primitive one of hoarding, will shift gradually to one of buying it for its intrinsic beauty and rarity (just as one buys diamonds). Furthermore, many millions of people in countries like India, who cannot afford any gold at all, may be able to buy some trinket for the first time. So gold buying will be more broadly spread among the population.

The prospect is, therefore, that not only can most newly mined gold in future be absorbed for fabrication, but that there will be a shortfall between supply and demand. In the next decade this gap is not likely to be more than two or three hundred tons in any year, which can be compensated for by gold winkled out of hoarding by higher prices (the French jewellery industry, for example, was already being supplied in 1970 and 1971 primarily by gold dis-hoarded in France). But in the long run, if the mining industry are right in their forecasts, the volume of newly mined gold must fall in the 1990s. The gap then between supply and demand might become substantial and the leeway could be made up either by dis-hoarding or by the gradual demonetisation of gold.

Which brings us back to the future of gold in the monetary system. The pattern there over the last few years is clear; gold takes up a smaller and smaller proportion of world reserves and the actual volume of gold in reserves has not increased for a decade. The outlook for the future also suggests that most, if not all newly mined gold, will continue to go into non-monetary hands. Only a major increase in the monetary price would really reverse this trend by reducing temporarily at least, private buying and possibly luring some gold out of hoards.

And this is where the debate about its future role comes in. What it really comes down to is whether or not one now regards gold as a commodity, just like silver or copper, which must stand on its own feet outside the monetary system, or as some sacred standard of value which must remain a cornerstone or the pivot of the monetary system.

Curiously enough the protagonists for and against line up very much in terms of age; those who want to increase the official gold price and maintain its monetary role are almost exclusively over fifty (and chiefly over seventy), while those under fifty (and certainly under forty) speak up for demonetisation.

The standard bearer for the gold standard is still Jacques Rueff, who came up in 1972 with a new twist to his old campaign to raise the gold price. Rueff's basic creed is that 'the only realistic international monetary system, which does not by itself generate inflation and deficit balance of payments, is convertibility into gold.'[1] So he has long pleaded for a doubling of the gold price coupled with full convertibility. But what about the bonus thus accruing to countries with large gold reserves, particularly in Western

[1] Jacques Rueff, speech to *The Financial Times* conference on Gold, London, October 1972.

Europe, which holds more than half of the world's monetary gold? This is where the new Rueff Mark Two plan comes in. 'If the price of gold were doubled,' he argues, 'the re-evaluation would unfreeze—on the figures of 31st December 1971—$10 billion in the U.S., $25,904 millions in Western Europe and $5,097 million in international institutions of "1934 dollars". The international agreement (to raise the price) would decide that the increment realised outside the United States—namely $31 billion—be offered to the U.S. through a long term loan—20 or 25 years—at a very low rate of interst. Such a loan joined to the $10 billion increment obtained from the U.S. gold reserve would permit the repayment of $41 billion of foreign claims.' Rueff sees his proposal as a 'Marshall Plan in reverse, a measure of fairness and gratitude in return for the generous contribution of the U.S. to the rehabilitation and reconstruction of the western world after the second world war.' Any statesman, Rueff concluded, who could bring about such a reconstruction of the international monetary system and restore convertibility 'would really save western civilisation.'

So far, no such saviour has stepped forward. And it is hardly likely that he will, for most politicians and bankers especially in the United States, would probably agree with the American senator who said that a return to the gold standard 'would be like repealing the twentieth century.'

Nowadays, moreover, we are not just considering gold's role in the twentieth century, but trying to plan its job in the twenty-first. There are a host of ideas on how gold may be sidelined or demonetised. Among their most lucid exponents is Fred Hirsch, a former member of *The Economist*'s staff, who was with the IMF in Washington from 1966 to 1972 working directly on plans to improve the international monetary system, and has since moved to Nuffield College, Oxford. Hirsch speaks with the advantage of someone with years of practical experience trying to manage our unruly monetary system, rather than as an academic toying with seemingly wise theories. Hirsch's basic standpoint is that 'two generations ago, gold became too valuable and too scarce for international money.'[1] He believes that gold should be phased out in favour of an international reserve system based on Special Drawing Rights. As a starter he would like to see the replacement of gold by SDRs in all IMF transactions and an ending to 'the taboo on official gold dealings in private markets associated with the two-tier system.' He also sees as desirable the 'expressing of currency values in terms of SDRs and dropping the SDR link to gold . . . and removing gold from the remaining various legal definitions of international monetary reserves, and also from official reserve statistics.'

But having said that, Hirsch recognises that many countries still have a strong attachment to gold; he sees no reason to deprive them of that pleasure. 'They should be fully free to do so, along with their other strategic and artistic reserves.'

These proposals do put gold on a more rational plain, while making allowances for those who still get emotional about it. Which was something that Lord Keynes always conceded, despite his dubbing gold a 'barbarous relic'. 'Gold,' he wrote, 'originally stationed in heaven with his consort silver, as Sun and Moon, having first doffed his sacred attributes and come down to earth as an autocrat, may next descend to the sober status of a constitutional king with a cabinet of banks; and it may never be necessary to proclaim a republic.'[2]

Today we have come much further along the road to proclaiming a republic, but the revolution will not be overnight. The emotions surrounding gold are still very

[1] Fred Hirsch, *The Politics of World Money*, special section in *The Economist*, 5 August 1972.
[2] J. M. Keynes, *Treatise on Money*, Macmillan, London 1930.

strong and to replace it as a standard of value is no easy task. As René Larre, general manager of the Bank for International Settlements stressed in his 1972 report, 'the essential role of gold at present is not its use as a means of settlement, but its use as the standard for the parities of currencies, and as the guarantee of value for the SDR itself and for creditor and debtor positions in the International Monetary Fund. It seems unlikely that the role of gold in these respects will be changed in the near future.'

Even the Americans, for all their talk about demonetising gold, are realistic enough not to push for it tomorrow. The current phrase of George Schultz, the American Treasury Secretary, in 1972 was that the United States looked for 'orderly procedures to facilitate a diminishing role of gold in international monetary affairs in the future.' The status of the SDR as the 'numaire' of the monetary system, Schultz added should be enhanced. While Dr. Arthur Burns, chairman of the U.S. Federal Reserve Board has admitted 'there is no chance in the world now of achieving agreement on the demonetisation of gold . . . (it) is not a practical possibility at this time.' 'At this time' is really the crucial phrase. The Americans and many other people, including most of the staff of the IMF, now assume that the eventual demonetisation of gold is inevitable. The argument is about the time scale. If silver is anything to go by, it could be a lengthy process. It is no easy matter to persuade people to give up gold as the basic standard of value and put their trust instead in some artificial creation like Special Drawing Rights. Furthermore, it is difficult to achieve while Russia and China are outsiders in the monetary system and are not members of the International Monetary Fund; they have no SDR allocations and it would be rather curious if the Western world switched to SDRs as a standard of value, while Russia and China adhered to gold. The Russians, as we have seen, have even spoken up in favour of gold as the basis for international trade. But international monetary co-operation has come a long way since the Bretton Woods conference in 1944; no doubt it will improve even more in the next quarter of a century, including bringing Russia and China in to the fold. Once they had access to IMF credits, gold would become more irrelevant. One could argue, therefore, that before the end of this century demonetisation of gold could proceed in relatively orderly fashion. By the 1990s, and certainly after the year 2000 the shortfall between gold production and the demand for the metal in jewellery and industry will be such that it can well be made up by gradually disposing of monetary gold on the commercial market. Even then many nations would probably choose to keep a healthy stockpile, just as they now keep emergency stocks of everything from uranium to copper and rubber.

The future of gold, therefore, is by no means gloomy, just because its status in the monetary system is on the wane. After all it still has plenty going for it; its intrinsic beauty, which originally recommended it to man, is quite untarnished and unchallenged. Diamonds may be a girl's best friend, but she is not likely to say 'no' to a gold bracelet either.

Appendix 1

GOLD PRODUCTION (NON-COMMUNIST COUNTRIES)

(in thousands of ounces fine)

	1966	1967	1968	1969	1970	1971	1972*
Republic of South Africa	30,880	30,530	31,085	31,165	32,145	31,370	29,215
Canada	3,275	2,965	2,690	2,545	2,410	2,210	2,080
U.S.A.	1,805	1,525	1,540	1,715	1,815	1,495	1,475
Australia	915	805	785	700	625	685	965
Ghana	685	765	725	710	705	695	730
Philippines	455	490	530	570	605	635	585
Rhodesia	550*	515*	515*	515*	500*	500*	500
Japan	255	255	240	245	255	255	240
Mexico	215	185	150	215	200	150	140
Colombia	280	255	240	220	205	190	185
Zaire	160	150	170	175	175	170	150
Nicaragua	145	140	140	110	115	105	115
India	120	105	115	95	105	115	110
Elsewhere	1,560	1,515	1,675	1,620	1,540	1,525	1,510
Total	41,300*	40,200*	40,600*	40,600	41,400*	40,100*	38,000
South Africa	74·8%	75·9%	76·6%	76·8%*	77·6%	78·2%	76·9%

*Estimated or provisional figures.
Sources: International Monetary Fund and Union Corporation estimates.

Appendix 2

FABRICATION OF GOLD IN NON-COMMUNIST COUNTRIES

Metric Tons

	1968	1969	1970	1971	1972
Europe	556·9	524·0	568·9	588·1	592
of which Italy	163·0	173·7	181·0	196·7	225
Germany	99·0	94·0	96·0	93·0	90
Spain	55·3	59·5	62·5	65·0	65
France	46·6	51·0	47·5	54·7	55
North & South America	317·8	320·2	287·1	314·1	304
of which U.S.A.	205·4	221·1	185·5	215·3	227
Brazil	46·2	36·0	32·7	34·0	23
Middle & Far East	346·1	329·3	471·0	427·6	292
of which India	140·1	108·3	217·0	179·0	107
Japan	53·0	65·0	74·0	80·6	96
Pakistan	30·0	20·0	30·0	25·0	15
Indonesia	20·0	25·0	30·0	25·0	10
Other	50·7	53·6	70·2	82·4	57
Total	1,271·5	1,227·1	1,397·2	1,412·2	1,245*

Reproduced by kind permission of Consolidated Gold Fields from *Gold 1972* by Peter Fells.* (1972 figures author's estimates.)

Appendix 3

GOLD FABRICATION BY END-USE

	1968 Metric Tons	%	1969 Metric Tons	%	1970 Metric Tons	%	1971 Metric Tons	%
Jewellery in advanced nations	441·6	34·7	458·9	37·4	448·6	32·1	495·7	35·1
Jewellery in developing nations*	485·6	38·2	454·1	37·0	614·0	43·9	562·6	39·8
Total Jewellery	927·2	72·9	913·0	74·4	1,062·6	76·0	1,058·3	74·9
Dentistry*	71·9	5·7	71·9	5·9	70·4	5·1	76·4	5·4
Electronics	84·2	6·6	102·6	8·4	93·6	6·7	91·3	6·5
Other Industrial and Decorative	60·9	4·8	69·1	5·6	69·1	4·9	74·9	5·3
Coins, Medals and Medallions	127·3	10·0	70·5	5·7	101·5	7·3	111·3	7·9
TOTAL	1,271·5	100·0	1,227·1	100·0	1,397·2	100·0	1,412·2	100·0

*About 10 tons of the gold attributed to jewellery usage in developing countries is probably used in dental applications.

Reproduced by kind permission of Consolidated Gold Fields from *Gold 1972* by Peter Fells.

Bibliography

This short bibliography is intended to lead the general reader toward a more detailed discussion of some aspects of gold. It is not by any means exhaustive. It is divided into four main categories: general interest books, the history of gold, the role of gold in the world's monetary system, and specialist topics.

GENERAL BOOKS

W. J. Busschau, *Measure of Gold,* Central News Agency, Johannesburg, 1949.

A. P. Cartwright, *The Gold Miners,* Purnell & Sons, Cape Town, 1962.

A. P. Cartwright, *Gold Paved the Way: The Story of the Gold Fields Group of Companies,* Macmillan, London, St. Martin's Press, New York, 1967.

A. P. Cartwright, *West Driefontein—Ordeal by Water,* Gold Fields of South Africa, 1969.

Paul Einzig, *Primitive Money,* Eyre and Spottiswoode, London, 1948.

Paul Ferris, *The City,* Gollancz, London, 1960, Pelican Books, 1962 and 1965.

E. J. Holmyard, *Alchemy,* Penguin Books, Harmondsworth 1957.

J. D. Littlepage and Demaree Bess, *In Search of Soviet Gold,* Harcourt Brace & Co., New York, 1938.

C. H. V. Sutherland, *Gold: Its Beauty, Power and Allure,* Thames and Hudson, London, 1959.

Francis Wilson, *Labour in the South African Gold Mines 1911–1969,* Cambridge University Press, 1972.

THE HISTORY OF GOLD

Leslie Aitchison, *A History of Metals*, 2 vols., Macdonald and Evans, London, 1960.

R. S. Anderson, *Australian Goldfields*, D. S. Ford, Sydney, 1956.

Pierre Berton, *The Golden Trail*, Macmillan, Toronto, 1954.

Sture Bolin, *State and Currency in the Roman Empire*, Almquist and Wiksell, Stockholm, 1958.

The Life of Benvenuto Cellini, written by himself, Phaidon Press Ltd., London, 1949.

Sir John Clapham, *The Bank of England*, 2 vols., University Press, Cambridge, 1944.

Edward Hyams and George Ordish, *The Last of the Incas*. Longmans Green, London, 1963.

W. P. Morrell, *The Gold Rushes*, A. & C. Black, London, 1940.

A. E. Murray, *Murray's Guide to the Gold Diggings*, D. S. Ford, Sydney, 1956.

Rodman W. Paul, *Californian Gold*, Harvard University Press, Cambridge, Mass., 1947.

P. D. Richards, *The Early History of Banking in England*.

THE MONETARY ROLE OF GOLD

W. J. Busschau, *Gold and International Liquidity*, South African Institute of International Affairs, Johannesburg, 1971.

Francis Cassell, *Gold or Credit*, Pall Mall Press, London, 1965.

Geoffrey Crowther, *An Outline of Money*, Thomas Nelson, London, 1948.

Paul Einzig, *The Destiny of Gold*, Macmillan, London, 1972.

Gold and World Monetary Problems, Proceedings of the National Industrial Conference Board Convocation, Tarrytown, New York, October, 1965, published by Macmillan, New York, Collier-Macmillan, London, 1966.

Sir Roy Harrod, *Reforming the World's Money,* Macmillan, London, St. Martin's Press, New York, 1965.

R. G. Hawtrey, *The Gold Standard in Theory and Practice,* Longmans Green, London 1947 (5th edition).

R. G. Hawtrey, *Currency and Credit,* Longmans Green, London, 1950 (4th Edition).

E. W. Kemmerer, *Gold and the Gold Standard,* McGraw-Hill, New York, 1944.

J. M. Keynes, *Treatise on Money,* Macmillan, London, 1930. Reissued 1950.

The Radcliffe Report on the Working of the Monetary System, Cmnd. 827, Stationery Office, London, 1959.

Robert Triffin, *Gold and the Dollar Crisis,* Yale University Press, New Haven, Conn., and London, 1971.

SPECIALIST TOPICS

Professor E. Bayer, 'Gold from the Oceans,' in *Chemistry,* Vol. 37, No. 10, October, 1964.

W. H. Emmons, *Gold Deposits of the World,* McGraw-Hill, New York, 1937.

Peter Fells, *Gold 1972,* Consolidated Gold Fields, London.

Gold—A world-wide survey, Economic Intelligence Unit, Charter Consolidated, London, 1969.

David Lloyd-Jacob & Peter Fells, *Gold 1969* and *Gold 1971,* Consolidated Gold Fields, London.

Edmund M. Wise, *Gold: Recovery, Properties and Applications,* D. Van Nostrand, Inc., New York, 1964.